FOREWORD

Since its inception in 1979, the Internal Combustion Engine Fall Technical Conference (FTC) has been focused on a pertinent theme selected for each year. These themes are based on those engine issues which are currently generating much technical activity. For the 1992 FTC, most of the papers deal with the theme "New Development in Off-Highway Engines."

The technical papers contained in this book were presented and discussed at the 1992 FTC, which was held October 4–7 in Waterloo, Iowa. These papers were grouped according to their specific topics into seven technical sessions. These sessions comprise:

- Mechanical Design
- Engine Environmental Issues
- Fuels, Combustion, and Emissions
- Engine Control Systems
- Engine Applications
- Simulation and Modeling
- Practical Application of Improved Quality Systems

This publication consists of 23 technical papers, produced by 57 authors.

A major objective of ASME is to develop and disseminate technical information that advances engineering technology. The Internal Combustion Engine Division continually strives to accomplish this objective as related to engine technology by sponsoring two conferences each year.

The published proceedings represent the main focus of the FTC. The conference also features other technical content worthy of special mention:

- An ASME Honda Lecture by Charles Amann, dealing with a pertinent engine subject, "Private Vehicles for Personal Transportation
- A free tutorial by Hans J. Bajaria which covers the features and application strategies for the new ISO 9000 quality standards
- A panel review of emissions regulations applicable to off-highway engines
- A session that presents practical application of improved quality systems
- A technical presentation competition among engineering students
- A tour through the John Deere Engine Works

The technical program for the 1992 FTC results from a team effort, involving many committed persons. We are especially appreciative for the efforts of the session organizers: Carl McClung, Terry Ullman, Reda Bata, Bruce Ingold, Jon Tice, Chuck Melcher, Rameshwar Sharma, and Joe Barcroft. Thank you to Charles Amann, Hans Bajaria, and the authors who shared their technical advancements with us.

We thank our conference host, the John Deere Power Systems Group, for their important contribution to the success of this conference. Thanks also go to the I.C. Engine Division Chairman, David Harrington, and the other members of the Executive Committee who provided their support.

Bruce Chrisman
Technical Program Chairman

ML

ICE-Vol. 18

NEW DEVELOPMENTS IN OFF-HIGHWAY ENGINES

presented at
THE 14TH ANNUAL FALL TECHNICAL CONFERENCE OF
THE ASME INTERNAL COMBUSTION ENGINE DIVISION
WATERLOO, IOWA
OCTOBER 4–7, 1992

sponsored by
THE INTERNAL COMBUSTION ENGINE DIVISION, ASME

edited by
BRUCE CHRISMAN
COOPER INDUSTRIES, AJAX-SUPERIOR DIVISION

THE AMERICAN SOCIETY OF MECHANICAL ENGINEERS
345 East 47th Street □ United Engineering Center □ New York, N.Y. 10017

Statement from By-Laws: The Society shall not be responsible for statements or opinions advanced in papers . . . or printed in its publications (7.1.3)

CONTENTS

1992 SOICHIRO HONDA LECTURER

Charles A. Amann
Principal Engineer, KAB Engineering
Bloomfield Hills, Michigan

THE LECTURER:

After a 42-year career with General Motors Research Institute, Charles Amann began his consulting firm, KAB Engineering. While at GM, he conducted research with automotive spark-ignition, diesel, gas turbine, steam and Stirling engines. As Head of the Engine Research Department for 16 years, he directed studies on engine combustion, emissions and fuel economy. He retired as Research Fellow and Director of the Engineering Research Council.

Mr. Amann has taught courses at Wayne State and the Universities of Minnesota and Arizona. He is a guest lecturer at Michigan State. He has 18 patents, is a frequent contributor to the technical literature, and has been an invited speaker in Europe and Asia as well as North America. He received the Woodbury Award for engineering management from the ICE Division of ASME, and the Outstanding Achievement Award from the University of Minnesota. He is a Fellow of SAE, a Member of ASME, and has been elected to the National Academy of Engineering.

THE HONDA LECTURE:

For 1992 the Honda Lecture deals with "Private Vehicles for Personal Transportation." It is included in this Proceedings Book.

The Soichiro Honda Lecture has been established as a National Lecture by ASME to recognize achievement and significant contribution in the field of personal transportation. Past recipients are Helmut List, President of AVL; Dr. Phillip Myers, Emeritus Professor, University of Wisconsin; Dr. Horst Hardenberg, Director of Advanced Truck Engine R&D, Daimler-Benz; Professor John B. Heywood, Director of the Sloan Automotive Laboratory and Professor of Mechanical Engineering at Massachusetts Institute of Technology; and Karl J. Springer, Vice President of the Automotive Products and Emissions Research Division of Southwest Research Institute.

PRIVATE VEHICLES FOR PERSONAL TRANSPORTATION

Charles A. Amann
KAB Engineering
Bloomfield Hills, Michigan

ABSTRACT

In the United States, private personal transportation has become dominated by the automobile, a platform supported on four wheels and propelled by an internal combustion engine (ICE). Some of the reasons why this combination has emerged as the preferred choice are reviewed. Since urban air quality has become an issue, the ICE has kept pace with progressively more stringent exhaust-emissions regulations. Future emissions standards will encourage the use of alternative fuels and battery-electric propulsion. Looking far into the future, the depletion of fossil-fuel resources and/or definitive evidence that the greenhouse gases are actually changing the global climate would foster a shift toward nuclear and solar energy. The automobile platform is compatible with such a shift. The ICE and the electric motor remain as potential motive sources, although they would face some difficult challenges.

INTRODUCTION

Personal transportation is one of the cornerstones of modern life. It is used to earn a living and to take a vacation, to shop for the necessities of life and to enjoy entertainment and recreation, to access doctor and barber, school and church, friends and relatives, and for myriad other purposes.

Personal transportation may be classified as either private or public. The latter category includes buses, trains and airplanes operated on schedules not under the direct control of the traveler. Private, rather than public, transportation accounts for the majority of personal trips made in the U.S. Vehicles for private transportation are the subject of this paper. The scope is further restricted to travel over land, thus excluding privately owned boats and aircraft.

The original mode of personal transportation was human locomotion. This imposed a speed-distance envelope on travel that depended on the physical condition of the traveler and the degree of exertion acceptable to him, the nature of the terrain, and the weather.

If a significant load was carried, the speed-distance tradeoff was affected adversely. There was also an upper limit to how much an individual could carry, independent of speed and

distance. Domesticating beasts of burden therefore represented a giant forward step in personal transportation.

A large variety of animal species has been enlisted to assist in personal transportation, but the one most common in the U.S. has been the horse. Some insight into the value of the horse in personal transportation is provided by Fig. 1. On this plot of average speed versus distance, dashed loci of constant trip time are superimposed. The "running" curve indicates current world-record performance for human athletes, a sort of upper limit for human locomotion. The generally negative slope of this envelope illustrates the drop in speed that accompanies fatigue. Below this running envelope is a second labeled "walking." It reflects the world-record performance of race walkers. This is still well above

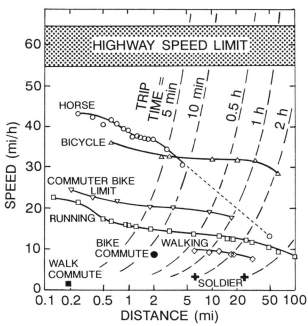

Fig. 1 - Speed versus distance for various means of personal transportation.

the typical human speed on foot. For example, the two "soldier" points mark military expectations for the trained World War II Expert Infantryman carrying a full field pack. The filled square is for the average person who walks to work, covering 0.19 mi at 1.4 mi/h (0.3 km at 2.25 km/h) [1].[1]

Well above the running-limit envelope is one for record race horses, extrapolated to a 50-mile point for horses traveling cross-country. As true for the human runner or walker, the average speed of the horse decreases with distance traveled. Even after discounting the record speeds on horseback for more typical travel by horse, that mode of transportation offers a substantial speed advantage over typical travel on foot. At the same time, travel by horse spares the human the physical exertion of travel on foot.

Crossing over the horse envelope in Fig. 1 is the record-performance curve for humans racing on bicycles. It is noteworthy that beyond a distance of about 5 mi (8 km), or a trip time of about 10 min, a world-class athlete on a bicycle can outspeed a horse. It is also evident that a human can more than double his best running speed with the help of a bicycle. Again, though, the bicycle is not normally driven at record speeds.

The bike "commuter limit" curve in Fig. 1 is the calculated limit for a healthy man [2] riding a commuter bike [3] while seated upright. Even this is well above what a commuter would willingly achieve in normal travel. For example, the average person biking to work travels 2 mi (3.2 km) at a speed of 9 mi/h (14.5 km/h), as represented on Fig. 1. It is nevertheless remarkable what that 19th century invention, the bicycle, can do for human travel capability, given a proper roadbed.

The shaded band at the top of Fig. 1 marks the legal highway speed limit in the U.S. of 55 to 65 mi/h (88 to 105 km/h). The modern automobile is quite capable of sustaining such speeds over long distances, without the droop characterizing the speed-distance envelope for human and animal propulsion. Thus on a proper roadbed, the automobile can travel faster and farther than these other options while carrying cargo loads, when required, that would be intolerable to a riding horse, or to a human either walking or riding a bicycle.

In the history of this nation, the speed characteristics portrayed in Fig. 1 have contributed to a shift from reliance on horses and bicycles for personal transportation to dependence on automobiles. At the turn of the century there were many more horses than cars, and there were over 200 bicycles built in the U.S. for every new automobile. That situation has long since been reversed, however. Taking the sum of the number of road horses and mules (i.e., excluding those dedicated to farming) and the number of automobiles as a crude measure of transportation mode [4], the transition is illustrated in Fig. 2, where percentage penetration of each mode is plotted against years.

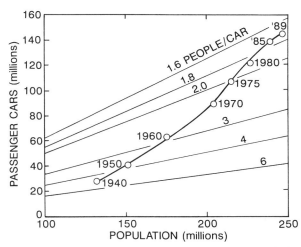

Fig. 3 - Historical increase in number of cars with population in the U.S.

As the horse was retired from significant use in private transportation, the number of passenger cars grew faster than the national population, as shown in Fig. 3. The points of cars versus population are labeled by year, and lines of constant people per car are superimposed. While there were nearly 5 people per car in 1940, that ratio dropped below 2 in the late 1970s and appears to be leveling off at about 1.7. A possible explanation for such asymptotic behavior has been advanced by Lave [5]. By expanding the vehicle count to include vans and trucks used in personal transportation and contracting the population count to exclude non-drivers, it appears that the asymptote about coincides with one personal-transport vehicle per licensed driver.

Thus, private personal transportation in the U.S. has become dominated by a vehicle platform supported on four wheels and propelled by an internal combustion engine. In this paper the rationale favoring the four-wheeled platform is first examined. Then the justification for choosing an internal combustion engine is reviewed. Finally, possible future propulsion systems for the era beyond fossil fuels are discussed.

THE VEHICLE PLATFORM

Important considerations in selecting the load-bearing platform of the vehicle include its support, stability, and directional control.

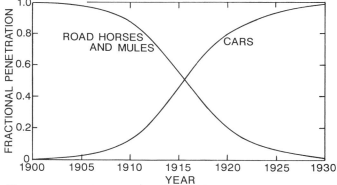

Fig. 2 - Historical transition from equine mode to automobile for private transportation in the U.S.

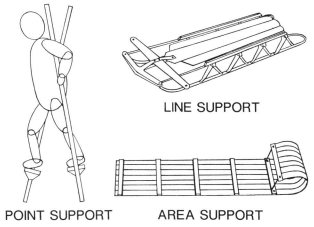

LINE SUPPORT

POINT SUPPORT   AREA SUPPORT

Fig. 4 - Examples of point, line, and area support of a load.

Platform Support

The contact footprint made on the ground may, in the limit, approach a point, a line, or an area. Examples are illustrated in Fig. 4. Point support is approached by stilts, line support by the runners of a sled, and area support by the undersurface of a toboggan.

In concept, the pneumatic tire of the automobile exemplifies point support. Neglecting the generally minor load-bearing contribution of the tire sidewalls, the footprint area of a pneumatic tire equals the product of its inflation pressure and the load it supports. Thus a 4000-lb (1816-kg) passenger car resting on tires inflated to 30 psi (207 kPa) creates on the ground a total footprint area of 0.96 sq ft (0.086 sq m). This is only about 1% of the plan area (length times width) of the vehicle platform.

The wheel is a unique point support because it allows forward motion of the vehicle with rolling rather than sliding friction. Thus the tractive effort required to transport a load is small compared to that with sliding-point support. Such support is exemplified by the travois of the American Indian, a load-supporting A-frame comprised of two poles, with the horse at the apex providing propulsion and the opposite ends of the poles dragging across the ground.

The point-support example of Fig. 4, stilts, avoids the sliding of the travois but would require a complex walking linkage for mechanization. Such mechanisms have severe speed limitations compared to the rotating wheel.

Line contact of such elements as the sled runner and the ice-skate blade are special cases of rubbing friction, made practical only because the interface with the ground is lubricated by the water on which the supports glide. Such runners and blades are unsuited for use on dry land.

The tracked vehicle, which in the limit approaches line support, is adapted to overland operation because, at least in forward motion, the track avoids sliding over the ground. As the track is widened to decrease ground pressure for better operation over unprepared terrain, it moves from the line-support concept toward the area-support concept. Indeed, a snowmobile with a track width approaching that of the platform can hardly be considered an example of line support.

The area-supported toboggan, like the sled and the ice skate, owes its utility to lubrication by snow and is not suited for general-purpose use. Satisfactory area support can be provided by using air as the lubricant, however. Such a vehicle may ride on either a cushion of air having a thickness measured in inches (tenths of meters) or a film of air having a thickness measured in thousandths of an inch (hundredths of a millimeter).

The Hovercraft concept, represented by the cross section of Fig. 5, provides an air cushion. An annular jet-sheet of air around the periphery of the platform is angled inward but must then deflect outward to escape the vehicle undersurface. By applying a radial-equilibrium analysis to the curved jet-sheet [6], it can be shown that with atmospheric pressure acting on the outside of the curved sheet (bend inner streamline), the pressure underneath the platform (bend outer streamline) must be above atmospheric. This super-atmospheric pressure under the platform is what supports it above the ground.

The Hovair principle [7] for creating an air film under the platform is illustrated in cross section in Fig. 6. For a circular platform, a circular plastic sheet is secured to the platform periphery. The center of the sheet is fastened to the center of the platform. One or more holes in the sheet, made near the center of the assembly, connect two air volumes. One is the cavity between the sheet upper surface and the platform above it. The other is the space between the sheet lower surface and the ground that is bounded by a ring of minimum ground clearance. Pressurized air delivered into the upper cavity inflates the sheet to a sort of toroidal shape, then passes through the hole(s) in the sheet to the volume

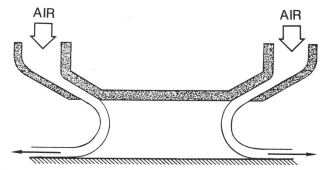

Fig. 5 - Area support with an air cushion.

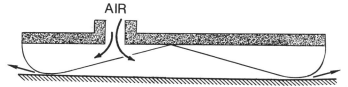

Fig. 6 - Area support with an air film.

beneath it, from which the air escapes radially through the minimum-clearance ring to provide the film of lubricating air.

Platform Stability

The static stability of a point-supported load depends on the number of support points. The unicycle and the pogo stick, two examples of single-point support, are statically unstable in all directions. To maintain either of these devices upright at a relatively fixed position on the ground, the operator must make continuous adjustments. Obviously, a single-wheeled vehicle is a poor choice for general transportation.

The stability of a two-wheeled vehicle depends on whether the wheels are located in tandem or in parallel planes. When the two wheels are in tandem, as exemplified by the bicycle, the vehicle is stable in pitch (longitudinal tipping). If properly designed, a bicycle ridden by an experienced operator is also stable in roll (lateral tipping) as long as the vehicle is moving forward at sufficient speed and the center of gravity is maintained in the plane of the wheels.

An important design factor in providing that stability is the trail, illustrated in Fig. 7. Trail is the horizontal dimension from the intersection of the steering axis with the ground back to the point

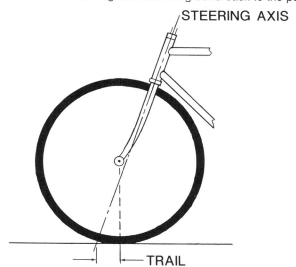

Fig. 7 - Trail dimension on a bicycle.

3

of tangency of the wheel with the ground. In the modern bicycle, trail generally ranges from 45 to 70 mm, with the larger dimension increasing high-speed stability at the expense of responsiveness when maneuvering [8]. The feasibility of riding with hands removed from the handlebars demonstrates the extent of lateral stability.

Vehicles supported on a pair of parallel wheels include the Roman chariot, the ox cart, and the rickshaw. These vehicles are stable in roll. However, they are unstable in pitch, as evidenced by use of the propulsion medium in these examples -- horse, ox, and human, respectively -- to maintain balance.

Static stability in both pitch and roll is possible with three supports, whether the pair of parallel wheels is followed by a single trailing wheel or preceded by the single wheel. Before World War II most airplanes had the single wheel (or alternatively, a tailskid) trailing, but modern airplanes usually have the single wheel leading, in the manner of a child's tricycle.

One problem encountered with four-point support that is avoided with three-point support is that arising from operating on a non-planar surface. Anyone who has experienced use of a four-legged table with one short leg can appreciate this problem. In the modern road vehicle it is accommodated by the spring element in the suspension system. This allows four-point ground contact when one of the wheels is located in a modest depression in the ground plane.

Another aspect of stability is the tolerance of the platform for asymmetric loading. In the case of the vehicle with two wheels in tandem, that tolerance is poor. If the center of gravity moves out of the plane of the wheels, the bicycle must be headed in the direction of the imbalance to avoid falling. The ability of the operator to steer a bicycle without using the handlebar merely by leaning in the direction of the turn demonstrates this coupling between imbalance and heading. That characteristic removes the tandem-wheeled platform from general-purpose platform support.

With wheels in parallel planes that characteristic is avoided. As illustrated in Fig. 8, an off-center load is balanced by asymmetric loading of the wheels. A platform supported on parallel tracks enjoys the same advantage.

When the load is supported on an air cushion, as also represented in Fig. 8, a problem arises. The resultant of the pressure force on the underside of the platform (P) acts through the center of the platform. Vectorially summing this pressure force with the off-center load (L) yields a transverse force (T) that leads to tilting of the platform. That transverse force exists because the horizontal momentum of the greater airflow exiting the high side exceeds that of the lesser amount escaping the low side. In the absence of frictional resistance, as would be provided by tires on a

wheeled vehicle, that transverse force causes the platform to move in the direction of the low side.

When an off-center load is applied to a platform supported by an air film under a flexible sheet, as illustrated in Fig. 8, the air-filled sheet assumes an asymmetric shape, moving the center of pressure of the footprint into alignment with the load. The degree of asymmetric loading tolerated by such a pad is severely restricted by the geometry of the flexible sheet. Consequently, the ability of three-point support to handle loading asymmetry can be utilized by supporting the platform on three pads, each one loaded centrally.

This solution is not free of problems, however. For any specified load, a given pad design functions satisfactorily only if the pressure of the air supplied to it is between two limiting values [7]. The lower bound is the floatation pressure necessary to lift the load, and the upper bound is a stability limit beyond which a vertical hopping motion is encountered. If two pads under a given platform receive significantly different loads, the air supply pressure required to float the heavier one may be above the stability limit for the lightly loaded one. For general-purpose use, therefore, a sophisticated automatic air-management system would be required.

Directional Control

In a platform supported on a pair of tandem wheels, or on three wheels or four, the driver normally controls the direction of travel by steering the front wheel(s). Although steering with the rear wheels is possible, the reason behind a preference for front-wheel steering is illustrated in Fig. 9. In driving a parked vehicle away from a parallel barrier like a curb or wall, rear-wheel steering invites a collision with the barrier.

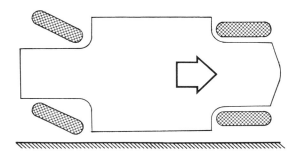

Fig. 9 -   Driving forward a vehicle with rear-wheel steering positioned parallel to a barrier.

The coupling between roll stability and direction of travel in the bicycle has already been noted. For three- and four-wheeled vehicles, these functions are decoupled.

A low center of gravity is desirable in vehicles supported on more than two wheels to avoid rollover accidents. For this and other reasons, the platform of the vehicle is typically supported at a distance less than one wheel diameter above the ground.

Steering such a four-wheeled vehicle with a solid front axle, reminiscent of a horse-drawn wagon of yesteryear as illustrated in Fig. 10, seriously depreciates the utility of the platform by occupying space that would otherwise be a load-carrying area. This shortcoming has been eased by the modern steering linkage, whereby each front wheel pivots on its own steering knuckle, also shown in Fig. 10. For the three-wheeled vehicle of Fig. 10, platform-loading flexibility is again sacrificed because the front wheel encroaches on space otherwise available for cargo.

Although the air-cushion supported platform avoids the issue of interference between wheels and load-carrying space, it carries an equivalent penalty in loading utility because of its voluminous

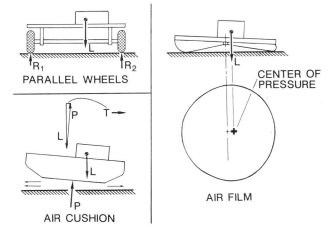

Fig. 8 -   Balancing an asymmetric load with parallel-wheel support, air-cushion support, and air-film support.

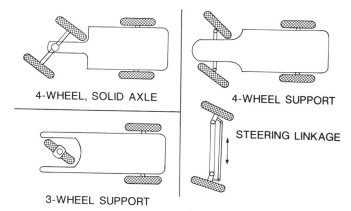

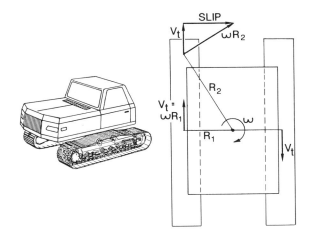

Fig. 10 -   Effect of steering on platform utility for (a) four-wheel support with a solid axle,(b) four-wheel support with a steering linkage, (c) three-wheel support.

Fig. 11 -   Pivoting a two-track vehicle, illustrating rubbing between track and ground plane.

air-handling system. This handicap is more severe in small vehicles the size of an automobile than in huge vehicles such as those employing this principle that are used to cross the English Channel. The air-film supported platform escapes this problem because of its much lower airflow requirement.

The turning of a two-tracked vehicle is represented in Fig. 11. The platform can be spun around its own center by turning the two tracks in opposite directions, as indicated. However, when the linear velocity of the center of the track at radius R_1 is compared to that near the end of the track, R_2, it is evident that considerable rubbing must take place between the track and the ground. Comparing such a vehicle to a tire-supported automobile in normal driving, it is realized that substantially greater wear due to rubbing would occur in the former than in the latter.

Steering a platform supported entirely on air must be accomplished either by aerodynamic surfaces, such as with a rudder, or by effecting a directional change in the momentum of an air stream, as by directing a jet normal to the direction of travel. The turning moment produced by a conventional rudder varies with the square of vehicle velocity, making the rudder ineffective at low vehicle speeds.

The shortcoming of aerodynamic steering of these types becomes apparent when control precision is considered. Two cars can pass each other traveling in opposite directions at a legal relative speed of over 100 mi/h (161 km/h) with a separation distance of only a car width, thanks to friction between tire and road. When that friction is eliminated by employing air support, a side gust of wind can easily blow the vehicle off course, and such close-proximity operation is out of the question.

To avoid this hazard, it has often been proposed to run high-speed air-supported vehicles in a guidance trough or some variation of this scheme. That is the equivalent of operation on a limited-access track and removes the air-supported vehicle from the field of private transportation.

Overall Assessment

Given the foregoing background, it is possible to assess the leading options discussed with respect to various qualities desired in a personal transportation vehicle. The prime candidates are listed across the top of Table 1. The qualities considered, listed vertically, are (a) static stability in pitch and roll, (b) the flexibility of the support system in accepting off-center platform loading, (c) the extent to which a platform of specified length and width can accept cargo without interference from the support system, (d) the ability to provide precise directional control without adverse effects on the support system and/or ground plane, and (e) the quality requirements of the road bed. This fifth characteristic has not been discussed, but it is obvious that a point-support device like the wheel that is intended to roll over the ground benefits from a reasonably smooth, hard roadbed. This restriction is eased for air-cushion support, but not for air-film support.

Reviewing Table 1, the most favorable options for general use appear to be either four wheels or two tracks. The tracked vehicle operates better on unprepared surfaces than the wheel but suffers from excessive rubbing between track and roadbed during turns.

Table 1

Comparison of Platform-Support Options
(OK = acceptable, X = deficient)

	Bicycle	Tricycle	Four-Wheeler	Two-Track	Air Cushion	Air Film
Static stability	X	OK	OK	OK	OK	OK
Loading flexibility	X	OK	OK	OK	X	X
Loading utility	X	X	OK	OK	X	OK
Directional control	OK	OK	OK	X	X	X
Roadbed requirement	X	X	X	OK	OK	X

Not brought out in this skeletal review is the noise that can arise with high-speed operation of a track.

The wheel is marked down for its roadbed requirement, but it has the advantage that a comprehensive network of prepared roads already exists in the U.S. One concludes from Table 1 that four-wheel support is the most sensible choice for a general-purpose personal private transportation vehicle.

Although the bicycle loses out in Table 1, its deficiency in roll stability is easily managed by an experienced operator. In remains a valid special-purpose option as long as the cargo is limited to the driver and a minimal additional load. The bicycle is not the vehicle of choice for hauling a steamer trunk, however. In addition, its comfort and safety, attributes not included in Table 1, are not competitive with those of a standard automobile.

To ease the propulsion task of the human-powered bicycle, it has been motorized into the moped, motorscooter, and motorcycle. Although these engine-powered two-wheelers are safer than the standard bicycle when mixed with automobile traffic on public roads, the risk in riding on them instead of in an automobile is still considerable. For example, Evans and Frick have reviewed U.S. accident statistics from 1975-89 and expressed the relative risk of driver fatality for mopeds (including motorscooters), motorcycles and four-wheeled vehicles of various weights in collisions with heavier vehicles [9]. In these terms, a relative risk of 85 for a motorcycle colliding with a medium-weight car means that the motorcyclist is 85 times as likely to suffer fatal injury as the driver of the car. The relative risk to drivers of mopeds, motorcycles, and light cars of 1014 kg mean weight is shown in Fig. 12 for two-vehicle collisions with heavier vehicles. In this breakdown, medium and heavy cars have mean weights of 1428 and 1833 kg, respectively. The comparative safety problem for operators of two-wheeled vehicles in mixed traffic is evident.

PLATFORM PROPULSION

Once a wheeled platform has been chosen for operation on a road network, there are four ways to propel it -- animal power, human power, jet momentum, and the driven wheel.

Animal Power

The historical rejection of the horse for widespread personal transportation has already been traced in Fig. 2. Given the expansion of personal transportation represented in Fig. 3, this switch in preferred transportation mode has been fortunate because it is unlikely that the environment could sustain a nation moved by horses today. If passenger cars were replaced one-for-one by horses, the resulting mass of solid waste requiring disposal would exceed a half billion tons per year, or over 350 g/mi. How fortunate that the automobile has rescued the U.S. from horse-based personal transportation.

Human Power

The bicycle is the most efficient way of applying human power to personal transportation. It is used for about half of the daily trips in Beijing, China, and in Delft, Netherlands, but only 8% of those in Manhattan, U.S.A. [10]. Nationwide, only about 1% of U.S. commuters travel to work on a bicycle or its motorized derivative, the motorcycle [1]. Bicycling remains popular as a sport in the U.S., but not for essential travel because of discomfort, the physical effort required, and the aforementioned concern about safety in mixed traffic on public roads.

However, the modern automobile has borrowed two of its important features from the bicycle technology that preceded it [11]. Dunlop, a Scottish veterinarian, popularized the pneumatic tire in Belfast in 1888. And in England, Starley applied a differential to a two-person tricycle with side-by-side seating. Starley did this to even the torque distribution between the two parallel driving wheels when the greater strength of his son, who pedaled one of the wheels while he pedaled the other, made it difficult to maintain a straight course.

Jet Momentum

Propulsive thrust is produced by accelerating a mass of air opposite the direction of travel. The turbojet and the engine-driven propeller are two common examples of thrust generation.

Propulsive efficiency is a measure of how effectively the kinetic energy generated in the jet stream is transferred to the vehicle. The inefficiency appears in the form of residual kinetic energy in the jet. As shown in Fig. 13, when the vehicle is stationary, the propulsive efficiency is zero because none of the jet energy is transferred to the vehicle. As the vehicle velocity approaches the jet velocity, the propulsive efficiency approaches 100%, but the mass airflow required to produce the thrust approaches infinity. This is illustrated by the second curve in Fig. 13, which shows the relative airflow required per unit of thrust force at a fixed vehicle speed. For high efficiency, therefore, one desires a small difference between jet and vehicle velocities and a large jet mass. For land-based vehicles this points to a large-diameter propeller rather than a turbojet.

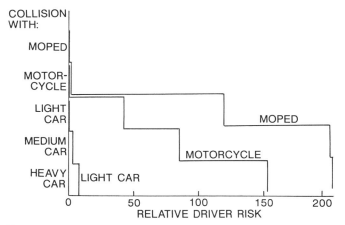

Fig. 12 - Driver-fatality risk in two-vehicle collision of moped, motorcycle, and light car with heavier vehicles listed at the left.

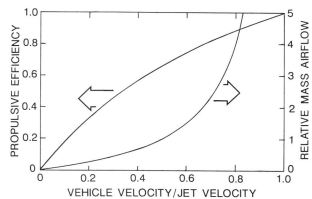

Fig. 13 - Effect of vehicle/jet velocity ratio on propulsive efficiency and on mass airflow required per unit of thrust developed.

The practical consequences of this can be visualized by imagining a queue of passenger cars awaiting a red traffic light to turn green. When that happens, the large volume of air to be moved and the associated noise would create an intolerable situation. Trying to back a wheeled vehicle with propeller propulsion into a tight curbside parking slot is equally taxing to the imagination. Automotive propulsion by jet momentum is feasible but hardly practical.

Driven Wheel

In the history of the passenger car, three means for powering its wheels have enjoyed significant market penetration -- the external-combustion engine, the electric motor, and the internal-combustion engine. Interestingly, all three saw concurrent use in automobiles at the turn of the century.

External-Combustion Engine - Since the external-combustion steam engine preceded the internal-combustion engine (ICE) in history, that it outnumbered the ICE in production cars of 1900 is no surprise. The steam engine enjoyed an insensitivity to fuel quality not shared by the ICE, but in a cold engine it took 20 to 45 minutes to raise enough steam pressure to drive.

The early steam engines had no condenser, requiring frequent stops for water. Later, the condenser was added. However, the closed cycle of the external-combustion steam engine demands that all of the cycle inefficiency be rejected to the coolant. (Note that because of the external combustion employed to deliver input heat to the closed thermodynamic cycle, additional heat is rejected in the burner exhaust stream, but that loss is not inherent to the closed thermodynamic cycle.) This contrasts to the open-cycle ICE, in which much of the cycle heat rejection exits with the exhaust gas. A result of this difference is that the steam condenser must be much larger than the radiator of an ICE of equal rated power. So demanding was the heat-rejection requirement of the steam engine that its condenser was typically inadequate for the job at high power output. Under these circumstances, some of the exhaust steam leaving the expander was released to the atmosphere to avoid a buildup of back pressure. Thus a tank of makeup water had to be carried on board and occasionally refilled along with the fuel tank.

In all the years the steam engine has existed, a suitable antifreeze for the water that is compatible with the temperatures encountered in the cycle has never been found. Hence, in winter climates the steam car could not be parked outdoors unattended for long periods without draining the water.

As the ICE improved, the steam engine became bulky and heavy by comparison, and its fuel economy was inferior. It was driven from the market by the ICE in the 1920s.

The late 1960s witnessed a renaissance of interest in the steam car, driven by the belief that its continuous combustion process promised lower exhaust emissions than possible with the intermittent combustion of the ICE. Two modern steam cars built in 1969 indicated that the promised emission advantage did not come automatically [12].

Upon starting these modern steam engines from room temperature, it took 1 to 2 min to raise steam pressure, but 2.5 to 6 min before the system was sufficiently warm to realize the performance potential of the engine. The fuel consumed during starting and warmup is a serious flaw present to some degree in all external-combustion engines. In the two steam cars cited, fuel spent during starting and warmup ranged from a quarter to a half gallon. To put this in context, a contemporary car with an urban fuel economy of 20 mi/gal consumes only 0.38 gal of fuel on the 7.5-mi Federal Urban Driving Schedule, including starting and warmup.

In addition to disappointing emissions and fuel economy, these steam engines showed poor power/weight and power/volume ratios, leading to substandard performance by current standards. Clearly, the ICE has passed by the steam engine, and the market-based decision of the 1920s favoring the ICE has not been revoked.

The Stirling engine is an external-combustion engine in which the preferred working fluid is high-pressure gaseous hydrogen rather than water/steam. The thermodynamic efficiency of the ideal Stirling cycle equals that of a Carnot cycle operating between the same temperature limits, so the Stirling engine promises greater efficiency than the steam-based Rankine cycle. The 1984 experimental evaluation of a car powered by a Stirling engine then under development revealed many of the same problems identified for the modern steam car, although generally to a lesser degree of severity [13]. Power/weight and power/volume ratios were low compared to the ICE, and urban fuel economy was not competitive, thanks largely to the fuel consumed during starting and warmup. Again the closed-cycle engine failed to catch the ICE, and there is no reason to expect that it will as a primary passenger-car powerplant.

Electric Motor - Second to the steam car in 1900 production was the battery-electric. It cost more to operate and had poorer performance than its competitors. However, it started promptly, unlike the steam engine, and it did not require hand cranking, as did the ICE of that day. It was quiet, in contrast to the ICE, and emitted no smelly fumes. In that age before the existence of a developed network of paved roads, its short range between battery chargings was less of an impediment than today. Similarly, the fact that it took all night to recharge batteries was not so serious a drawback.

Since 1900 the ICE acquired an electric starter, silenced its exhaust effectively with a muffler, and improved its combustion to make exhaust gases more tolerable. The road network improved in both mileage and quality, leading to longer trips. By the 1920s, the battery-electric car joined the steam car in extinction.

Internal-Combustion Engine - Reasons that the ICE displaced competing propulsion options during the first quarter of this century are clear from the above discussion. The early spark-ignition ICE was troubled by the phenomenon of combustion knock. As knowledge of combustion grew, however, that problem was made manageable, largely through the introduction of leaded gasoline, improvements to petroleum refining, and better combustion-chamber design.

In the early 1950s a new issue arose -- the contribution of the engine to air quality. The decade of the 1970s was greeted with nationwide tailpipe emission standards. Successively tighter standards have led to positive crankcase ventilation, exhaust-gas recirculation, the catalytic converter, the air pump, the evaporative canister, electronic controls, and replacement of the traditional carburetor with fuel injection. Given current tailpipe standards, it takes 25 contemporary car engines to emit as much tailpipe hydrocarbon (HC) and carbon monoxide (CO) as did a single car engine from the precontrol era. At the same time, the emission of nitrogen oxides (NOx) has been decreased 75%.

New federal standards will be phased in beginning in 1994. In recognition of the lower reactivity of methane in the chemistry of photochemical smog, that hydrocarbon will no longer be counted. The new non-methane hydrocarbon (NMHC) standard is only 61% of the present total HC standard, and the NOx standard will be decreased by 60%.

In California, which has the poorest air quality in the nation, even stricter standards are coming. That state has developed a successively cleaner series of low-emission vehicle standards that culminate in the ZEV (Zero-Emission Vehicle). ZEV is currently a synonym for an electric vehicle (EV). With no tailpipe emissions and no hydrocarbon fuel stored on board, the EV is a ZEV locally. However, given the origins of U.S. electricity, stack emissions at central power stations dictate that the EV is not a ZEV on a regional basis. California requires that 2% of the sales of major manufacturers be ZEVs in 1998, that fraction escalating to 10% in 2003.

Federal law prohibits 50 different sets of emission standards for 50 different states; each state must accept either the federal regulation or the California standards. Many states having trouble with air quality, mostly in the northeast, have either opted for or are considering adopting California standards. If those states all accept the California standards, close to half of the new-car production could be affected. As the law now stands, that would force a considerable number of EVs onto the road.

In addition to lowering emission standards further and regulating gasoline composition in cities with the worst air quality, the federal Clean Air Act Amendment of 1990 calls for some use of alternative fuels by the 1996 model year. Included are methanol (usually blended with 15% gasoline to make M85), natural gas, reformulated gasoline, electricity, and hydrogen. With the exception of electricity, all of these are compatible with the ICE. The first four options are likely to receive the greatest attention in the ICE community.

Atmospheric ozone concentration is the normal index of urban smog. In recognition of the fact that not all HC species have the same ozone-forming potential (OZP), California has set out to assign an OZP value to each of the over 150 different species found in ICE exhaust. This will lead to a set of reactivity adjustment factors (RAF) for each alternative fuel when used in each of the California low-emission-vehicle categories. The measured tailpipe non-methane organic gases (NMOG = HC - methane + oxygenated HC species like aldehyde) will then be multiplied by the appropriate RAF to adjust the NMOG for regulatory purposes. This technique should help the alternative fuels to meet future California tailpipe standards.

The most stringent California emission standards short of the ZEV are those of the ULEV (Ultra-Low Emission Vehicle). Its requirements are 0.04/1.7/0.2/0.008 g/mi NMOG/CO/NOx/formaldehyde. With the help of alternative fuels, including reformulated gasoline, the ULEV standards are not necessarily beyond the potential of the ICE, although they certainly involve a high degree of uncertainty. The ICE/EV hybrid would incur less risk of compliance. Given the anticipated slow penetration of the pure EV, the ICE will continue to play a strong role in personal transportation well into the 21st century, certainly as the vehicle prime mover and perhaps contributing a share of the propulsion duty through its application to hybrids.

Torque Characteristics - One important trait of the prime mover used to drive the wheels of a road vehicle is its full-load torque characteristic, for it is the difference between the torque delivered to the driving wheels and the road-load torque requirement of the vehicle that establishes its ability to climb hills, and to accelerate. In Fig. 14 typical torque curves, normalized to the torque at maximum power, are plotted against the fraction of maximum speed for a DC motor, a steam engine, and a spark-ignition engine.

In the upper part of its speed range, field current is controlled at fixed armature voltage and the electric motor is capable of producing the hyperbolic torque curve of constant power shown in Fig. 14. At low speeds its torque is normally restricted to an essentially constant value to avoid overheating the motor, with armature voltage being varied at full field strength. This is the general shape of an ideal torque curve for vehicle propulsion because if the power were held constant over the complete range of speeds, the torque would be infinite at zero speed. The tires cannot accept torques approaching infinity at low vehicle speeds without slipping.

The steam-engine torque curve is for a four-cylinder uniflow expander with a poppet intake valve in the head and exhaust ports in the cylinder walls [14]. It rises continuously toward a maximum at zero engine speed. Although this is a well behaved torque

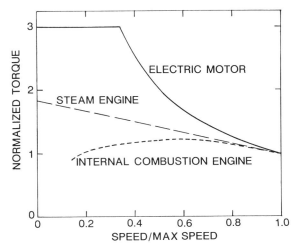

Fig. 14 - Torque characteristics of a DC electric motor, an external-combustion engine, and an internal-combustion engine.

curve for vehicle propulsion, the torque rise from rated to zero speed is sufficiently small that a transmission is required behind the engine to provide adequate performance. By stepping down the rotational speed of the driveshaft in its lowest gears, the transmission multiplies the torque delivered to the wheels at low vehicle speeds.

The ICE is seen to have the poorest torque curve. It extrapolates to zero at some low engine speed. This is because the torque developed depends on the mass of air trapped in the cylinder during each cycle, and with typical valve timing, the engine cylinder becomes an ineffective air pump at low speeds. Actually, rough running disallows operation in this regime. In contrast, the steam engine escapes this fate because it draws its working fluid from what amounts to a reservoir of high-pressure steam in the steam generator, which provides a supply not directly linked to expander speed. Given this inability to operate at engine speeds approaching zero, the ICE must be coupled to a transmission that includes some form of slipping element for accommodating the speed difference between an idling engine and a stationary driveshaft. That difference is normally handled with a clutch in manual transmissions or a torque converter in automatics.

Efficiency Characteristics - A second characteristic of importance in the wheel-driving system is its energy efficiency. In a car it is difficult for an external-combustion engine to achieve fuel economy competitive with that of the ICE for several reasons.

First, the external-combustions engine normally operates on a closed cycle, which requires the heat generated in combustion to be transferred across the physical boundary of a heat exchanger. As previously noted, this first requires energy to be deposited in an initially cold heat exchanger before the working fluid becomes hot enough to perform significant work, and that stored heat is dissipated on shutdown.

Second, efficiency depends on both the maximum and minimum temperatures of the working fluid in the cycle. The maximum cycle temperature is restricted to a level lower than the temperature limit of the material in the heat exchanger that separates the combustion flame from the working fluid. Similarly, the minimum cycle temperature must be greater than ambient because of the temperature drop across the cycle cooler, which is the condenser in the steam engine. That elevation of minimum cycle temperature above ambient is minimized by enlarging the cooler, but there is little room for that in a passenger car.

Third, energy is discharged from the external-combustion path in the form of exhaust gas at above ambient temperature. This represents an energy loss that is not included in the normally calculated thermodynamic-cycle efficiency.

Fourth, the auxiliary load is normally greater than for an ICE engine, including, as it does, a combustion-air blower and cooling fan(s). The cooling fan(s) must be of larger capacity than for an ICE radiator because of the larger heat rejection attending a closed cycle, which lacks an exhaust heat sink.

Comparing the energy efficiencies of EVs to production cars with ICEs is always awkward because the vehicles are not equal. The production car is capable of cross-country driving; the near-term EV is not. Nevertheless, attempts at such comparisons can be instructive.

Wang and Deluchi have tabulated test results in city driving for a dozen different EVs [15]. The energy drawn from the battery per lb of vehicle test weight for each mile traveled is plotted against test weight in Fig. 15. Also shown are data points from the energy consumptions of the most efficient 1991 production cars with automatic transmissions in various EPA test-weight classes as tested on the FUDS (Federal Urban Driving Schedule). These points were calculated from the measured urban mi/gal after a 10% depreciation for real-world (as opposed to chassis-dynamometer) driving. The slopes of the two lines are 74 and 450 mW-h/lb-mi, respectively. But ratioing these coefficients does not imply that the EV is six times as efficient as the production car.

The EV encounters a number of energy losses between the battery terminals and the fossil-fuel energy entering the central power station, be it in the form of coal, natural gas, or oil. The energy chain is illustrated in Fig. 16. The current average efficiency for the conversion of fossil fuel energy to electricity is about 33% [15]. Typically, another 8% is lost in delivering the electricity to the wall plug. Next comes the losses in the battery charger and in the battery during both charging and discharging. The average ratio of energy extracted from the battery to energy supplied from the wall plug has been measured at 0.67 for 15 test points in a car with a lead-acid battery [16] and separately reported as 0.624 for tests of five different EVs [17]. Combining these typical efficiencies from Fig. 16, about 5 kW-h must be supplied into the central station for every kW-h extracted from the battery.

Carrying the gasoline chain back to the corresponding point, i.e., crude oil entering the refinery, refining and distribution involves about a 10% energy loss.

Applying these factors, the energy required becomes 370 mW-h/lb-mi for the EV and 500 mW-h/lb-mi for the car with a gasoline ICE. Again, though, the ratio of these coefficients does not provide

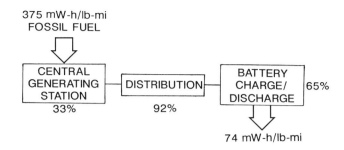

Fig. 16 -   Typical energy losses from fossil fuel input to battery output for battery-electric car.

a proper comparison. In a sample of eight EVs with room-temperature batteries, the battery weight averaged nearly 30% of the vehicle test weight [17]. In an ICE car, the weight of the filled fuel tank is typically less than 5% of the vehicle test weight. Therefore, comparing the two types of cars at equal test weights is unrealistic.

Wang and Deluchi matched eleven EVs using room-temperature batteries with their ICE-powered counterparts, finding that, primarily as a result of battery weight, the EV averaged 43% heavier [15]. Dividing the 500 mW-h/lb-mi for the ICE car by 1.4 to put it on the same car-size basis as the EV, the ICE coefficient becomes essentially equal to that of the EV. In other words, there is no great efficiency difference in city driving between the contemporary ICE car and the contemporary EV of comparable size. Of course, an EV weighing 40% more than an ICE car of equal size remains inferior in payload capacity. Moreover, the ICE car still offers better performance, faster replenishment of energy when the on-board energy supply has been expended, and an ability to make cross-country trips that is denied the EV.

Emission Characteristics - A third important trait of prime movers is their emission contribution. When the steam engine received its second wind in the 1970s, the reasoning was that its steady combustion should make emissions control easier than in the ICE. In 1975, however, the catalytic converter was added to the ICE. Because it can remove over 90% of the regulated emissions from the exhaust stream, that 1970's argument has lost much of its appeal.

The emissions associated with the EV depend on the source of electricity. If it is generated from water or nuclear power, then the EV is indeed clean. In the U.S., however, 54% of the electricity currently comes from coal combustion, 4% from oil, and 9% from gas [18]. Using central-station emission values from EPRI [18], it is possible to calculate the g/mi of VOC (Volatile Organic Compounds), CO, NOx, and sulfur dioxide associated with a 4200-lb EV, which might be considered a replacement for a 3000-lb ICE car. In Table 2, the results are compared to the current federal standards.

It is apparent from Table 2 that the EV offers substantial benefits in VOC and CO in trade for a modest increase in NOx. Sulfur dioxide emissions from gasoline-fueled ICEs are sufficiently low that there has been no need to regulate them, so clearly the EV would increase that contribution to acid rain substantially. Additionally, the particulate emissions associated with an EV operated on coal-based electricity would increase markedly compared to a gasoline-fueled ICE car.

EPRI projects that post-1995 powerplants, NOx will be reduced by a factor of 2 to 3.5 for all three fossil fuels listed in Table 2. That still leaves the EV above the Tier I standard of 0.4 g/mi that will be in effect federally for passenger cars in that time frame. In addition, a nearly fivefold reduction in sulfur dioxide is anticipated for the post-1995 coal plant [18].

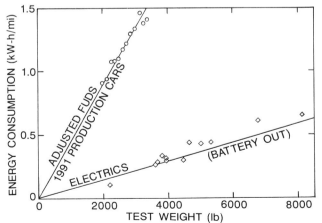

Fig. 15 -   Required energy in city driving for (a) battery-electric (energy out of battery) and (b) production cars with automatic transmissions (energy in gasoline).

Table 2

1989 Emissions of a 4200-lb Electric Vehicle
(g/mi in urban driving)

	VOC	CO	NOx	SO_2
1989 standards	0.41*	3.4	1.0	--
EV/coal	<0.01	0.06	1.84	4.81
EV/oil	0.03	0.10	1.05	2.64
EV/gas	0.01	0.10	1.21	<0.01

*0.41 g/mi total hydrocarbons

Alternative ICEs - The earlier discussion of ICEs focused on the spark-ignition engine because it has so dominated automotive propulsion for the past 70 years. Two alternative ICEs that deserve mention are the compression-ignition diesel and the continuous-combustion regenerative gas turbine.

When it comes to distance covered on a gallon of fuel, the diesel engine is the uncontested winner. First, it gains about a 13% advantage over the gasoline ICE because of the greater energy content of a gallon of diesel fuel compared to gasoline. The diffusion combustion of its fuel spray avoids the knock problem accompanying the premixed combustion of the traditional spark-ignition ICE, so the diesel is free to use a higher compression ratio. Third, diffusion combustion allows load to be controlled by varying the overall fuel-air ratio in the cylinder, thus avoiding the part-load pumping loss of the traditional spark-ignition ICE during the intake stroke.

The diesel automobile first appeared on the market over 50 years ago. More popular in Europe than in the U.S., its market penetration in the U.S. peaked at 6% in 1981 and is now less than 1%. The comparatively low cost of automotive fuel in the U.S. and the unfavorable price differential between gasoline and diesel fuel have discouraged sale of the more expensive diesel engine in U.S. passenger cars.

Despite keen federal interest in higher automotive fuel economy, evolving emissions regulations have done little to encourage continued development of the passenger-car diesel by U.S. companies. The diesel has found a niche market in light- and medium-duty trucks and vans, however, some of which are used in personal transportation.

The diesel continues strong in the heavy-duty truck market. The most challenging emissions regulations for all classes of diesel have been those for NOx and particulate matter. Technologies being pursued in the heavy-duty segment to satisfy more stringent standards include improved turbocharging and aftercooling, higher injection pressures, and increased flexibility in fuel injection, both its timing and its rate characteristic.

Particulate traps are being explored to satisfy the regulation of that emission, but cost and durability experience have been disappointing. Less costly routes to lower particulates include decreased oil consumption, reformulated fuel, and use of an oxidizing catalytic converter to decrease the soluble fraction of the exhaust particulate matter. Unfortunately, nearly all such measures increase the cost disadvantage of the diesel relative to the spark-ignition ICE.

A high-risk version of the conventional diesel that is being researched is the LHR (Low Heat Rejection) diesel, sometimes inappropriately called the "adiabatic" diesel [19]. In this concept, the traditional liquid-cooling system is eliminated, necessitating the use of ceramics for the walls of the cylinder and the combustion chamber to accommodate the increased temperature that results. The hotter walls lead to a reduction in volumetric efficiency, so turbocharging must be incorporated to compensate for the lost power.

Thermodynamics dictates that only a small fraction of the coolant heat rejection thus eliminated can be converted directly to crankshaft work. Most of the energy conserved by eliminating the coolant appears in the exhaust in the form of increased gas temperature. To capitalize on this energy, the LHR diesel is normally compounded, i.e., a second turbine geared to the crankshaft is added downstream of the turbocharger turbine. Such a compounding system is quite ineffective at light loads, so the LHR diesel appears poorly suited to the passenger car.

Other impediments to acceptance are compliance with stringent NOx emission standards, identification of a lubricant that will withstand the hotter cylinder-wall temperature, and development of structural ceramics or ceramic coatings combining increased reliability with acceptable cost. The uncooled LHR diesel will not appear in personal-transportation vehicles until it has been commercialized in the heavy-duty automotive application, if ever.

Another high-risk ICE, being supported by the federal government, is the regenerative gas turbine. Both two-shaft and single-shaft versions are being explored. The single-shaft gas turbine is unsatisfactory for automotive use unless coupled with a continuously variable transmission, parallel development of which appears to be unsupported.

The two-shaft automotive gas turbine has 40 years of development history behind it. This is approximately twice the length of time it took the spark-ignition ICE to dislodge the steam car and the battery-electric from the marketplace. Still the first commercially viable automotive gas turbine has yet to appear.

The two-shaft gas turbine enjoys smooth, vibration-free operation. It has an admirable torque characteristic, although the rise in full-load torque with decreasing output speed is still insufficient to avoid the need for at least two transmission-gear ratios. Principal barriers for early passenger-car gas turbines were non-competitive fuel economy, slow acceleration response, and excessive NOx emissions.

The new thrust in automotive gas turbines has been toward use of structural ceramics to allow a higher turbine inlet temperature. The lower density of structural ceramics, compared to high-temperature metallic alloys, promises some help with acceleration response. Higher limiting temperature increases cycle efficiency, promotes increased NOx emissions, and decreases engine size by boosting the power delivered per unit mass of airflow.

course, burning biomass without replacing it, as occurs in the clearing of rain forests, does increase the greenhouse-gas inventory.

Biomass/ICE

Ethanol is made in the U.S. today from corn, but the process is too expensive to make ethanol a competitive primary fuel. The total energy available form U.S. corn-based ethanol is also small compared to the gasoline-energy consumption of the U.S. fleet. These two characteristics presently relegate corn-based ethanol to its role as an environmentally acceptable octane-enhancing additive in gasohol.

Methanol is the current alcohol of choice for the U.S. fleet. The present direction in the U.S. is toward M85, a blend containing 15% gasoline. Although methanol can be made from wood, current resources and economics point toward methanol from natural gas.

Whether the alcohol used is ethanol or methanol, shortened range is a disadvantage. In Fig. 19, power density is plotted against energy density for a variety of options. Power density is an indicator of performance, energy density of range. The gasoline bubble is a reference point for a contemporary automobile, based on the power and range <u>after</u> the gasoline energy has undergone conversion in the powertrain, and debiting the fuel by the mass of the powertrain needed to effect that conversion [23].

The ethanol and methanol bubbles are for the same-displacement engine and the same size of fuel tank. Power is up slightly, an advantage of alcohol fuels. However, range is down substantially, primarily because the energy content of a gallon of alcohol is approximately 2/3 and 1/2 that of a gallon of gasoline for ethanol and methanol, respectively. To minimize the range penalty, ethanol is obviously preferred over methanol.

The National Renewable Energy Laboratory is developing a process to convert cellulose into ethanol. If the process evolves satisfactorily, it has been estimated that enough ethanol might be produced from domestic sources at a reasonable cost to satisfy the current automotive fleet [24], as illustrated in Fig. 20, where recent national energy consumption for non-petroleum and petroleum resources are represented by the two bars on the left. The unshaded portion of the center bar denotes transportation energy. Most of it goes to automobiles, trucks and buses, as indicated. The balance is used in airplanes, railroads, ships, and other miscellaneous applications.

The bar at the right in Fig. 20 shows an estimate of the cellulose-based ethanol potential from (a) municipal, agricultural and forest wastes, (b) wood from forestland, (c) crops grown on land now idled by federal policy, and (d) new cropland [24].

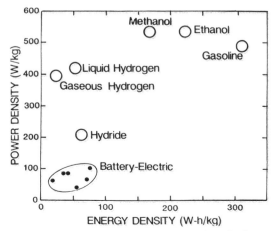

Fig. 19 - Power density versus energy density for various automotive propulsion systems.

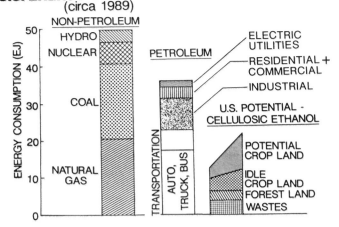

Fig. 20 - U.S. energy consumption from non-petroleum sources (left), petroleum sources (center), and estimated potential for cellulosic ethanol (right).

If the ethanol projections prove to be true, solar/biomass fuel could go a long way toward meeting the energy demand for transportation. Unfortunately, this route does not free the nation totally from current exhaust-emission concerns.

Hydrogen/ICE

If the ICE is fueled with hydrogen from electrolysis via either nuclear or solar electricity, the tailpipe emissions of HC and CO are virtually eliminated because the lubricating oil becomes the only source of carbon. NOx emissions would require control, however. The ready ignitability of hydrogen makes the engine prone to backfiring, but ways exist to manage that phenomenon.

The biggest problem with hydrogen is that of on-board storage. The severity of this problem is indicated by the low energy density of the hydrogen bubble in Fig. 19, which represents a volume of gaseous hydrogen stored at 3000 psi (20 MPa) in the space normally occupied by a gasoline tank. Power is down relative to gasoline because of the space occupied in the intake manifold by the hydrogen gas.

Liquid hydrogen (LH$_2$) appears to be much better than gaseous hydrogen in Fig. 19, but it is still a poor match for gasoline. It must be kept at 20 deg K (-423 F) to avoid boiloff. The release of vapor that is bound to occur creates a safety concern in enclosed spaces like garages. Another drawback to liquid hydrogen is its unsuitability for pipeline transmission.

Hydrogen is storable in a metal hydride, where it is bound in the metal and released upon heating. Conceptually, that heat is available from the engine exhaust gas. Hydride storage involves a large weight penalty, however, which contributes to its poor showing in Fig. 19. The hydrides most effective in storing hydrogen require a temperature source hotter than engine exhaust gas under many driving conditions.

Battery/EV

In the first quarter of this century, the ICE car drove the battery-EV from the marketplace because the EV had (a) poor range, coupled with a long recharge time, (b) poor performance because of its low power/weight ratio, and (c) high cost associated with limited battery life. The prior discussion of contemporary EVs suggested that these problems have not disappeared. Also, the EV showed no particular efficiency advantage. As for emissions, it claimed a definite advantage on a local basis, but on a regional basis it seemed merely to trade a reduction in some emissions for an increase in others.

If the electricity comes exclusively from a nuclear source, the central-station efficiency concern is diminished because no fossil fuel is consumed. Emissions, both local and regional, are eliminated in exchange for a radioactive-waste disposal problem. The other EV problems listed remain as technical challenges, primarily associated with the battery.

One of the challenges is system weight. To decrease the weight of the motor, its speed can be increased. The DC motor has the excellent torque characteristic illustrated in Fig. 14. However, as shown in the schematic of Fig. 21, it relies on brushes for commutation, and brushes wear. The speed of the DC motor is limited to keep brush maintenance at an acceptable level.

Solid-state electronics enable use of the AC induction motor, also diagramed in Fig. 21, for traction purposes. This option avoids brushes, facilitating a higher speed for reduced size and weight. As suggested in Fig. 22, when such a motor is operated at constant frequency f_1, its torque rises rapidly from zero at the synchronous (no slip) speed to a maximum with some modest degree of slip. The preferred operating point lies between the synchronous speed and the peak-torque speed. By employing a modulating inverter, voltage and frequency can be changed to provide curves f_2, f_3 and f_4. The electronic control system can manage frequency and voltage to provide the solid torque-speed curve of Fig. 22, which resembles the curve for the DC motor shown in Fig. 14.

The brushless permanent-magnet motor provides another high-speed option. Also shown schematically in Fig. 21, it, too, uses a variable-frequency inverter-type controller to match the needs of the vehicle.

By far the heaviest element in the battery-electric system is the battery pack, not the motor and controller. The gravity of the situation is illustrated by the battery-electric patch in Fig. 19, calculated for a variety of advanced batteries (at 50% depth of discharge, circa 1988) in combination with the corresponding motor and controller.

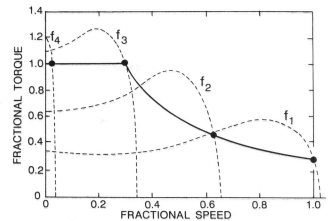

Fig. 22 - Securing an ideal torque-speed characteristic from an AC induction motor via solid-state control of frequency and voltage.

Great improvements have been made in the utility of the lead-acid battery that powered the EV of yesteryear. Sealed maintenance-free lead-acid batteries are now commonplace, and in some designs the liquid electrolyte has been replaced by a gel. Despite this progress, the lead-acid battery lags behind in energy density compared to some of the advanced batteries under development.

This is illustrated in Fig. 23, where approximate bands of power density versus energy density are plotted for three general classes of battery in their current state of development. Lines of constant discharge time are superimposed. Using the right boundary of the sodium-sulfur band as an example, it is seen that if energy is extracted from the battery at a rate of 37 W/kg, the battery charge will last for 3 h, supplying 117 W-h/kg. If energy is extracted at a higher rate of 97 W/kg, however, the battery is discharged in 1 h while supplying only 98 W-h/kg. If the energy were stored instead as a volume of liquid fuel, the total energy available would be independent of its rate of usage.

The Pb band in Fig. 23 is for various versions of the lead-acid battery. The Ni band includes Ni-Zn, Ni-Fe, and Ni-Cd pairs. These nickel batteries offer a better performance/range tradeoff than the lead-acid type, but are not without shortcomings. The nickel electrode is significantly more expensive than one made of lead. The Ni-Zn battery falls short of the lead-acid on cycle life, i.e., the number of discharges before requiring replacement. The Ni-Fe

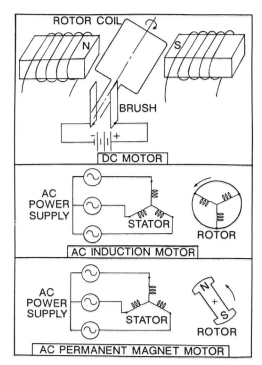

Fig. 21 - Schematics of three types of electric motor suitable for tractive use.

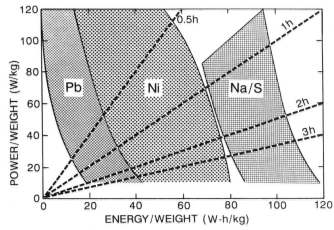

Fig. 23 - General range of power density versus energy density for various battery classes.

11. Wilson, S. S., "Bicycle Technology," Scientific American, Vol. 228, Nov 3, 1973, pp. 81-91.

12. Vickers, P. T., Mondt, J. R., Haverdink, W. H., and Wade, W. R., "General Motors' Steam Powered Passenger Cars - Emissions, Fuel Economy and Performance," SAE Trans., Vol. 79, 1970, pp. 2099-2120.

13. Haverdink, W. H., Heffner, F. E., and Amann, C. A., "Assessment of an Experimental Stirling-Engine-Powered Automobile," Proceedings of the 22nd Automotive Technology Development Contractors' Coordination Meeting, Dept of Energy, pp. 151-166, 1984.

14. Amann, C. A., Sheridan, D. C., Sagi, C. J., and Skellenger, G. D., "The Uniflow Steam Expander -- Its Relation to Efficiency of the SE-101 Powerplant," IECEC Proceedings, 1972, pp. 960-970.

15. Wang, Q., and Deluchi, M. A., "Comparison of Primary Energy Consumption by Gasoline Internal Combustion Engine Vehicles and Electric Vehicles," Paper 910804, Transportation Research Board 70th Annual Meeting, Jan 1991.

16. Conover, R. C., Hardy, K. S., and Sandberg, J. J., "Vehicle Testing of Near-Term Batteries," SAE Paper 800201, 1980.

17. Deluchi, M. A., Wang, Q., and Sperling, D., "Electric Vehicles: Performance, Life-Cycle Costs, Emissions, and Recharging Requirements," Transportation Research A, Vol. 23, No. 3, 1989, pp. 225-278.

18. Electric Power Research Institute, "Electric Van and Gasoline Van Emissions: A Comparison," EPRI TB.CU.177.10.89, 1989.

19. Amann, C. A., "Promises and Challenges of the Low-Heat-Rejection Diesel," ASME Trans., J. of Engineering for Gas Turbines and Power, Vol. 110, July 1988, pp. 475-481.

20. Collman, J. S., Amann, C. A., Matthews, C. C., Stettler, R. J., and Verkamp, F. J., "The GT-225 -- An Engine for Passenger-Car Gas-Turbine Research," SAE Trans., Vol. 84, 1975, pp. 690-712.

21. Murphy, T. E., "Power System Optimization for Passenger Cars," SAE Paper 850030, 1985.

22. Dietrich, W. R., "Criteria of Future Fuels for Stationary Gas and Diesel Engine Drives," Engine and Environment: Which Fuel for the Future?, AVL List GmbH, Graz, Austria, 1991, pp. 221-241.

23. Amann, C. A., "The Passenger Car and the Greenhouse Effect," SAE Trans. Vol. 99, Sect. 6, 1990, pp. 1646-1665.

24. Lynd, L. R., Cushman, J. H., Nichols, R. J., and Wyman, C. E., "Fuel Ethanol from Cellulosic Biomass," Science, Vol. 251, No. 4999, 1991, pp. 1318-1323.

25. Burke, A. F., "Hybrid/Electric Vehicle Design Options and Evaluations," SAE Paper 920447, 1992.

26. Krumpelt, M., and Christianson, C. C., "An Assessment and Comparison of Fuel Cells for Transportation Applications," Argonne National Laboratory ANL-89/28, 1989.

ICE-Vol.18, New Developments in Off-Highway Engines
ASME 1992

AN OPPOSED PISTON DIESEL ENGINE

Joey K. Parker, Stuart R. Bell, and David M. Davis
Department of Mechanical Engineering
University of Alabama
Tuscaloosa, Alabama

ABSTRACT

Typical conventional diesel engine designs are based on arrangements of single piston and cylinder sets placed sequentially either inline or offset ("V") along the crankshaft. The development of other engines, such as the opposed piston type, has been motivated by potential advantages seen in such designs which may not be viable in conventional inline or V engine arrangements. Several alternatives to conventional engine design have been investigated in the past and some aspects of these designs have been utilized by engine manufacturers.

The design and development of a proof-of-concept opposed piston diesel engine is summarized in this paper. An overview of opposed piston engines is presented from early developments to current designs. The engine developed in this work is a two stroke and uses four pistons which move in two parallel cylinders that straddle a single crankshaft. A prechamber equipped with a single fuel injector connects the two cylinders forming a single combustion chamber.

The methodology of the engine development process is discussed along with details of component design. Experimental evaluations of the assembled proof-of-concept engine were used for determining feasibility of the design concept. An electric dynamometer was used to motor the engine and for loading purposes. The dynamometer is instrumented for monitoring both speed and torque. Engine parameters which were measured include air flow rate, fuel consumption rate, inlet and outlet coolant temperatures, coolant flow rate, inlet air and exhaust temperatures, and instantaneous cylinder gas pressure as a function of crank position. The results of several testing runs are presented and discussed.

BACKGROUND

An opposed-piston engine is one in which two pistons oppose each other in each cylinder, each piston receiving the combustion forces and transmitting those forces to an output shaft, see Figure 1. The method of coupling the forces to a common output shaft has been a major difference among the many designs attempted. There are several distinct benefits in the basic design of opposed-piston engines which have caused a great deal of interest in their development. They are typically very well balanced, mostly due to the symmetry of design which establishes similar masses moving in opposite directions at all times. By incorporating piston-controlled intake and exhaust ports, the opposed-piston design uses fewer moving parts than conventional engines while providing an excellent means of uniflow scavenging. The total stroke of an opposed-piston engine is divided between two pistons; thus, piston speed and piston and cylinder wear are greatly reduced. Engine life is, therefore, increased while maintenance efforts are decreased. These and other advantages make the opposed-piston more efficient than its single-piston counterpart, both thermodynamically and mechanically, and thus opposed-piston engines are often smaller and lighter than conventional engines with comparable power ratings.

Although still considered a novel concept, the opposed-piston principle for diesel engines was introduced not long after the invention of the diesel engine itself. Opposed pistons were first used experimentally in early steam engines, and then in several gas engines beginning in 1874 [1]. Several inventors patented opposed-piston gas engines in England during the 1880's, but Wilhelm von Oechelhauser and Hugo Junkers are credited for the original development of these engines due to extensive work done at the Experimental Station for Gas Engines in Dessau. There, in 1892, they developed the first two-stroke opposed-piston engine. Oechelhauser and Junkers also worked independently on opposed-piston engines, resulting in the patent of the Oechelhauser gas engine in 1896 and of the Junkers gas engine in 1901 [1]. The concept was quickly applied to diesel engines by several researchers, including Junkers.

In 1907 Junkers developed his first opposed-piston oil engine using the design of the original Oechelhauser and Junkers gas engine patented in 1892. After several major changes, the standard Junkers design was established and was used extensively by Junkers and others in the years that followed [1]. Although others were unsuccessful with this design, Junkers continued to improve and produce opposed-piston engines, and in 1950 the Junkers "Jumo 205E," the powerplant of many WWII German aircraft, had the highest thermal loading on the piston of any engine in production [2,3].

Inspired by moderate success in the automotive and aircraft industries during the early 1900's by Gobron Brillie', Faccioli, Lucas, and Junkers, British marine engineers began to develop large diesel

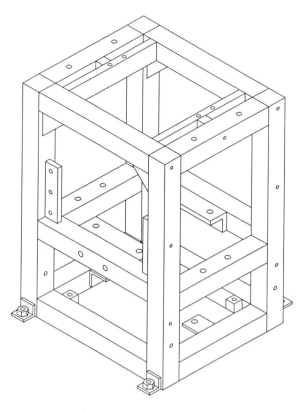

Figure 3: Open engine frame.

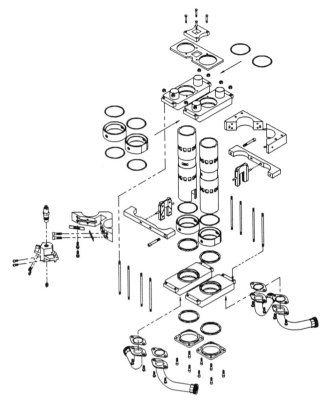

Figure 4: Exploded view of H-4 engine.

by a General Electric dynamometer. It is important to note that the current design of the engine was intended to prove the concept, not to develop a production or even prototype opposed piston engine.

The prechamber is removable and connects tangentially to each of the liners through two short transfer ports. The upper two pistons connect to an upper bridge (item 2 in Figure 2) and the bridge to the center connecting rod (item 1 in Figure 2) which transfers the combustion force to the upper crankrod and crankshaft. Two sets of bearing supports are used to constrain the movement of the center connecting rod. The lower set of pistons are connected to a lower bridge which is also constrained to vertical movement using a linear bearing arrangement. The lower bridge is connected to the crankshaft using two crankrods which connect to the two outside crank throws straddling the center crankrod and throw. The cylinder liners were manufactured from steel tubing bored to match the pistons which were obtained from a commercial vendor. Intake ports were machined at the upper midsection of the liners and exhaust ports at the lower. Intake and exhaust manifolds slip over the liners for routing the intake air and exhaust. The engine is water cooled using two slip-on water jackets which cover the outside portion of the liner between the intake and exhaust manifolds. The prechamber fits along the center of the liners. Figure 4 shows an exploded view of the major components connecting to the liners.

The engine design offers several potential attributes over conventional inline or V engine designs which served as motivation for the work. First, the opposed piston design should lead to lower net loads on the crankshaft. The axial loads exerted by the upper piston set due to gas pressure are counteracted by the axial loads from the lower piston set. Inertial loads due to the rotating and translating masses of the upper and lower piston sets are also cancelled to some degree. The crankshaft bearings can, therefore, be physically smaller than for a conventional engine with similarly sized

pistons. This leads to lower weight and less material cost for the engine block. Smaller crankshaft loads could also lead to less bearing wear, which reduces maintenance requirements.

Vibration forces generated by the engine should also be minimized when compared to conventional engines of the same size. If the lag angle between primary crankshaft throws is 180 degrees, then complete balancing of all pressure and inertial loads would be possible. The "proof-of-concept" engine currently uses a 150 degree lag angle to facilitate scavenging of the exhaust products from the cylinder. Therefore, the primary and secondary shaking forces are not completely cancelled, but can be reduced over conventional engine characteristics.

The side or thrust loading on the individual pistons of the engine are reduced by several orders of magnitude. The piston skirts are not required to act as bearing surfaces as in conventional engines. A much lighter weight piston could likely be used with greatly reduced wear on the pistons and rings.

The engine design has eliminated the cylinder head and associated valve equipment for a reduction in moving mechanisms to accomplish scavenging. The elimination of valves could lead to lowered maintenance requirements and reduced cost of manufacturing and assembly. The lack of a cylinder head has a potential benefit with respect to thermal performance as the surface area bounding the working gas is reduced. Improved thermal efficiency can translate to lowered fuel consumption for a given engine.

The scavenging process for the engine is a uniflow design which is also desirable. Gas enters the cylinder through the upper intake ports and exits through the lower exhaust ports. Although most other opposed engine designs also share some of these advantages, they commonly use two crankshafts or are coupled by complicated external gearing or pushrod arrangements.

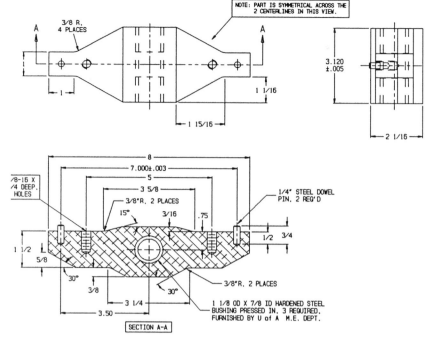

Figure 5: H-4 lower bridge.

The minimal configuration for the H-4 engine uses a single fuel injector to feed four pistons. This reduces the cost and complexity of the fuel injection system for the engine. The large piston-to-injector ratio would also be advantageous in potential special applications, such as dual-fuel engines.

Initial testing of the engine identified excessive bending (buckling) in the lower piston rod which was used as a linear bearing as well. A new lower piston rod with approximately the same weight but 2-3 times the bending strength was designed and machined. Also, a new one-piece lower bridge was designed to reduce weight and accept the new lower piston rods, see Figure 5. The horizontal bearing surfaces for the lower bridge were moved from the piston rods to the central pins of the bridge. Better lubrication of the bearing and more precise positioning at the point of force generation were distinct improvements. The lower bridge pins accepted two bronze bearing pads which were positioned between hardened steel plates forming the linear bearing for the lower bridge. This lower bearing was force lubricated and was rigidly attached to the engine frame and crankshaft bearing supports.

FACILITY DESCRIPTION

The H-4 engine was set up for testing in the Internal Combustion Engines Laboratory at The University of Alabama. Figure 6 is a schematic of the experimental arrangement, illustrating the instrumentation pertinent to the engine testing presented in this paper. As shown, the engine was attached to an electric dynamometer, which was equipped with a scale to measure the loading force of the dynamometer, thus providing a means of calculating the brake engine torque produced. Engine speed was also monitored using a speed transducer located on the opposite end of the dynamometer.

The intake system included a surge drum and a roots type blower to provide air to the inlet manifolds. An ASME long radius nozzle was used in the inlet of the surge drum to obtain air flowrates. The pressure drop across the nozzle was measured with a U-tube manometer and a calibration curve was used to convert the pressure drop to air flowrate. The fuel flowrate measurement was obtained by recording the time for consumption of a given amount of fuel through a small buret which was graduated in milliliters.

The fuel injector was fitted with a pintle lift indicator in order to monitor injection timing. A pressure transducer was installed in the prechamber through a port opposite the injector. The transducer allows measurement of the combustion chamber pressure, which allows more in-depth study of the combustion event. An encoder attached to the fuel pump shaft, which operates at engine speed, was used to record crankshaft position so that the pintle lift and cylinder pressure could be referenced to crankangle locations. Data from the pintle lift indicator, the pressure transducer, and the encoder were recorded using a digital storage oscilloscope. This data provided a means by which to explore the time history of various parameters within the combustion chamber, including pressure, heat release rates, injection timing, and combustion duration.

TESTING RESULTS

An initial firing test was conducted after all component upgrades and motoring tests were completed. The engine was motored to a speed of approximately 500 RPM by the dynamometer before diesel injection began. After the rack was engaged the engine began to run smoothly and accelerated to a speed of about 700-750 RPM before the fueling rate was decreased. The indicated power produced by the engine at this speed was in the range of 7-8 hp, which is approximately the same as the friction power.

Close inspection of the engine after the initial firing test uncovered two problems. The lower bridge used two cantilever pins to attach the connecting rods of the crankshaft. This design was not adequate and allowed the hardened steel bushings to separate from the aluminum block. Also, the addition of the hardened steel bushings left an inadequate cross-section of aluminum to withstand the bending load applied as the engine fired. Consequently a small amount of bending was detected in the lower bridge. These two

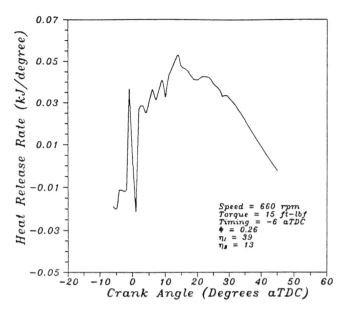

Figure 13: Heat release rate as a function of crank angle for torque of 15 ft-lbf.

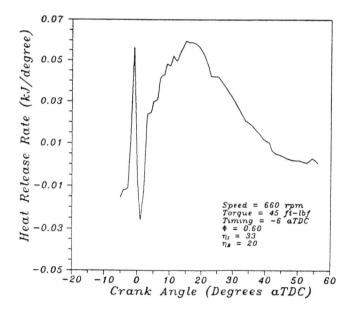

Figure 14: Heat release rate as a function of crank angle for torque of 45 ft-lbf.

REFERENCES

1. Wilson, W. Ker, "The History of the Opposed-Piston Marine Oil Engine," *The Institute of Marine Engineers Transactions,* v.58, Oct.-Nov., 1946.

2. Allen, O.F. , *The Modern Diesel,* Prentice-Hall, N.Y., 1947.

3. Middleton, J.H.D., "The High-Speed Light-Weight Diesel Engine," *The Engineer,* v.190, Nov. 17, 1950.

4. "The Opposed-Piston Principle for High-Power Diesel Engines," *Power,* v.57, Jan. 9, 1923.

5. "Features of Diesel Engines With Opposed Pistons," *Power Plant Engineering,* v.45, May, 1941.

6. "Crankless Opposed-Piston Diesel Engine," *Power,* v.78, Feb., 1934.

7. Ruskin, P., "Novel Design Features of Fairbanks-Morse Opposed-Piston Engines," *Automotive and Aviation Industries,* v.92, Jun. 15, 1945.

8. "F-M's Small OP Engine," *Diesel Power and Diesel Transportation.* v.31, May, 1953.

9. "Turbocharging The O-P Engine," *Diesel Power,* v.36, Mar., 1958.

10. "New Opposed-Piston Diesel," *Mechanical Engineering,* v.87, Jun., 1965.

11. "Eccentric-Type Opposed-Piston Marine Diesel Engine," *Engineering,* v.169, Jun. 9, 1950.

12. Smyth, R. and F.J. Wallace, "Comparative Performance Assessment of Various Compression-Ignition Engine Configurations in Combination with Compressors and Turbines," *Proc. Instn. Mechanical Engineers,* v.181, 1966-67.

13. Lapeyre, J.M., "Internal Combustion Engines," United States Patent 4,543,917

14. Lapeyre, J.M., "Internal Combustion Engine," United States Patent 4,566,408

15. Lapeyre, J.M., "Internal Combustion Engine," United States Patent 4,732,115

16. Bell, S. R., J. K. Parker, and J. M. Lapeyre, "Development of a Proof-of-Concept Opposed Piston Engine," *ASME Paper 90-ICE-7,* presented at the Energy-sources Technology Conference & Exhibition, January 1990, New Orleans

COMBINED AIR-OIL COOLING ON A SUPERCHARGED
TC & IC TAM DIESEL ENGINE

Ferdinand Trenc
TAM Research and Development Institute
Maribor, Slovenija

Radislav Pavletič
Department of Mechanical Engineering
University of Ljubljana
Ljubljana, Slovenia

ABSTRACT

In order to reduce maximum cylinder wall temperatures of an air - cooled TC&IC Diesel engine, large longitudinal and circumferential temperature gradients, a curved, square cross-section channel supplied with engine lubrication oil was introduced into the upper part of the cylinder wall. Numerical analyses of the heat transfer within the baseline air-cooled cylinder and intensive experimental work greatly helped to understand the temperature situation in the cylinder at diverse engine running conditions. The results of the combined cooling were greatly affected by the design, dimensions, position of the channel and the distribution of the cooling oil flow and are presented in the paper.

DESCRIPTION OF THE PROBLEM

In order to establish the problem of inadequate temperature distribution in an air-cooled cylinder and consequently to solve it, analyses of the baseline cylinder temperature distribution has to be completed first. The baseline cylinder was therefore equipped with more than 70 temperature sensors - thermocouples. Some temperatures served as temperature boundary conditions and some of them to check the accuracy of the numerical model of the heat transfer within the cylinder wall.

The position of the particular temperature sensors in the baseline cylinder wall is shown on Fig.1; thermocouples defining the internal cylinder wall temperatures were distanced 1.2 mm from the cylinder working surface. Basic measurements were performed on a single and four cylinder TC & IC air-cooled Diesel engine.at diverse engine running conditions: at well stabilized loads, speeds and temperatures. Results of measurements showed pronounced assymetric circumferential temperature distribution, relatively high maximum cylinder temperatures and high longitudinal temperature gradients of the cylinder inner wall. Temperature assymmetry is mainly provoked by the influence of the intake and exhaust channel, by the

shortages of the cooling surfaces design, and the physical properties of the cooling air. Fig.2 shows circumferential wall temperature distribution for diverse cylinder cross sections at full load and peak torque engine speed (according to the symbols in Fig.1). Critical transverse section I, that corresponds to the top position of the first piston ring (when the piston reaches its TDC), determines most of the limiting temperature values of the cylinder wall. Circumferential assymetry of temperature is defined as the biggest temparature difference in the particular cross- section and reaches its critical peak value of nearly 35°C in section I. Wall temperature distribution in the longitudinal cross- section "3" (the hottest) and "8" (the coolest) are typical too; maximum average longitudinal temperature gradients in the upper part of the cylinder reach almost 1°C/mm. The influence of the entire cooling fin insulation of the lower half of the cylinder is interesting; wall temparatures in that region increase by approx. 6°C only, whereas the top cylinder area temperatures remain unchanged (Fig.2).

On-field application of an engine is mostly characterized by sudden and frequent changes of speed and load. According to the results of the experiments performed, only a very thin layer of the cylinder wall reacts simultaneously - due to the thermal capacity of the cylinder material - to sudden changes of the engine thermal load.The "information" of higher thermal load propagates slowly throughout the cylinder wall until it reaches the cooling fins. During the period of the thermal stabilisation, local temparature differences in the cylinder can reach even higher values compared to those at stabilized conditions and the highest loads. Local temperature differences can provoke additional thermal stresses and deformations, and can therefore influence the reliability and longevity of the engine. Some typical engine load-time simulations have been performed on the laboratory test stand to determine their influence on the time temperature distribution in the cylinder wall. Temparatures at diverse places were measured simultaneously throughout the

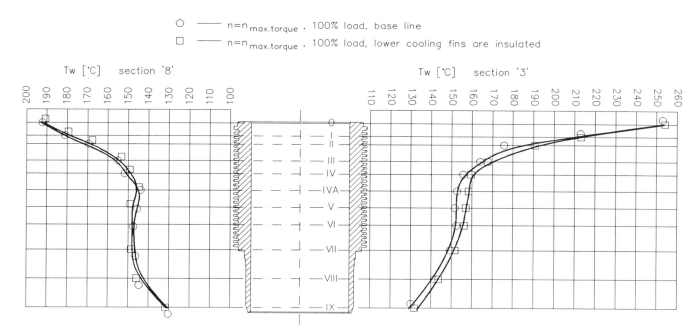

FIG.3 LONGITUDINAL CYLINDER WALL TEMPERATURE DISTRIBUTION; INFLUENCE OF THE PARTIAL COOLING SURFACE INSULATION

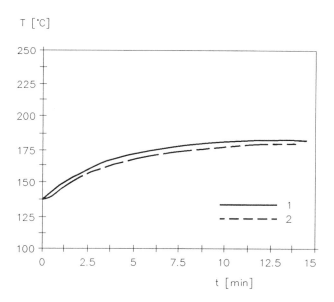

FIG.4 TEMPERATURE TIME DISTRIBUTION OF A POINT IN THE CYLINDER WALL; 1- MEASURED, 2- CALCULATED

Taking into account all the above mentioned, one can conclude, that thermal problems of an air-cooled cylinder involve only a small and limited portion of the cylinder wall surface: from the top (contact surface with the cylinder head) to the transverse section "I" (less than 25% of the cylinder hight!).

SOLUTION OF THE PROBLEM AND THE RESULTS

According to the results in chapter 2, a curved, square cross-section and horizontal channel supplied with lubricating oil was introduced into the upper part of the cylinder wall. Dimensions and position of the channel consequently followed the results of the analyses of the baseline air-cooled cylinder and the analyses of the heat transfer problems in laminar flows of viscous fluids within the short, curved channels reported by Trenc (1992).

Diverse forms of uniform oil flow and split flow combined cooling systems were tested on a single-cylinder engine. Detailed analysis of the heat transfer in the channel, based on statistical treatment of the experimental data showed that local values of the heat transfer coefficient varied intensively along the branches of the channel (Trenc, 1992). Experiments were performed at diverse engine running conditions and diverse cooling oil flows. As the result of multiple regression, a non dimensional mathematical model for the heat transfer coefficient, depending on the Reynolds (Dean), Prandtl and Grashof number was established (Trenc, 1992). Variation of the local heat transfer coefficient can be used to suit local demands for supplementary cooling and can therefore help to balance unequal circumferential temperature distribution in the cylinder wall. Fig.6 shows the distribution of the average (for the four channel walls) local Nusselt Nuz numbers.

Local oil temperatures for the two channel branches are presented as well; oil temperature increase of only 6°C was observed despite the high engine loads. On the other hand, no local overheating of the cooling oil was noticed. Results of the combined air-oil cooling are presented on Fig.7 and 8. Fig.7 shows the influence of the combined cooling on the temperature distribution in the inner cylinder wall at the typical engine loads.

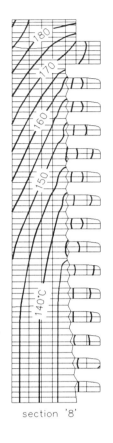
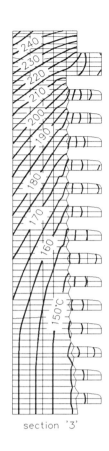

section '8'   section '3'

FIG.5 CALCULATED TEMPERATURE DISTRIBUTION IN
TWO LONGITUDINAL CYLINDER SECTIONS

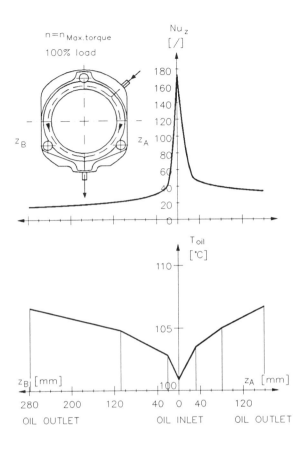

FIG.6 AVERAGE LOCAL NUSSELT NUMBER NUZ AND
COOLING OIL TEMPERATURE DISTRIBUTION FOR
TWO BRANCHES OF THE OIL CHANNEL

Comparison was made for air and combined cooled cylinder. A pronounced assymetry of the polar temperature distribution in the air cooled cylinder was substantially levelled in the combined cooled cylinder to only 10°C. Maximum wall temperatures in the transverse section "I" of the combined cooled cylinder never exceeded 180° C even at 10% overload (specific power corresponds to 24kW/l); their level corresponded to the partial (67%) load temperature level of the baseline cylinder. Similar or better results were obtained at the engine rated speed. Axial cylinder wall temperatures for the baseline, combined and water cooled cylinders are presented in Fig.8. Temperature gradients are reduced for the combined cooled cylinder (see the enclosed table) and the circumferential temperature assymetry is even less pronounced when compared with the water cooled version. Temperatures in the lower half of the combined cooled cylinder are substantially higher compared to those of the water cooled one, but not too high (corrosion problem due to the water vapour condensation can be better controlled).

Supplementary oil cooling is self controlled; higher engine loads result in higher average cooling oil temperatures. As the result, higher heat transfer coefficient values are obtained and more intensive cooling tends to decrease cylinder wall temperatures. There is an optimum quantity of the oil required for the combined

cooling cylinder: 6 to 10kg/min are suitable for the rated specific power of 22 to 25 kW/l engine swept volume. Influence of the cooling oil flow on the circumferential wall temperature distribution is presented in Fig.9. Besides, cooling oil inlet - outlet temperature increase and maximum cylinder wall temperature differences in the transverse section "I" are presented in the enclosed table. Although optimum oil flow was substantially increased, no adequate wall temperature reduction was observed.

EXPERIMENTAL METHODS

Experiments performed were basically linked to the engine performance data at diverse running conditions. Measurements on the single and four cylinder engines were performed on a fully electronically controlled test stand and supported with AVL - PUMA 4 control system. Simultaneous temperature measurements were performed with a digital temperature data logger and with a Data acquisition system HP 3052A.

0.2 mm diameter NiCr - Ni thermocouples were used for temperature measurements; the whole measurement chain was calibrated by known temperature properties of the Pb-Sn binary alloy.

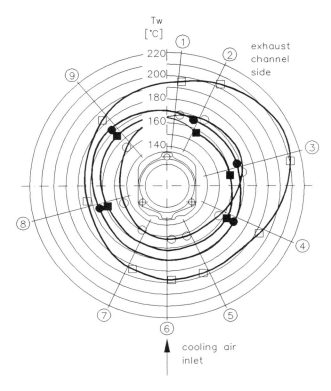

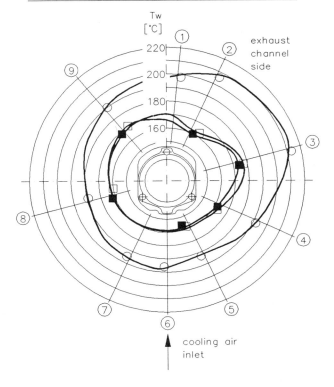

cooling		engine load [%]
○	—— air	67
□	—— air	100
■	····· combined	100
●	—— combined	110

$n = n_{max.torque}$
100% engine load

cooling		$\Delta T_{oil\ in-out}$ [°C]	$\dot{m}_{oil}$ [%]	$\Delta T_{cyl.}$ [°C]
○	—— air	/	/	/
□	—— combined	7.6	60	19
■	—— combined	6.6	100	17

FIG.7 CIRCUMFERENTIAL TEMPERATURE
DISTRIBUTION OF THE CYLINDER WALL FOR AIR
AND COMBINED COOLED ENGINE

FIG.9 INFLUENCE OF THE OIL MASS FLOW ON THE
CIRCUMFERENCIAL TEMPERATURE DISTRIBUTION
FOR AIR AND COMBINED COOLED CYLINDERS

DISCUSSION

High temperatures and temperature gradients within the upper part of the air cooled cylinder do not tolerate any increase of the engine specific power. Axial and circumferential wall temperature gradients can provoke additional thermal stresses and deformations of the cylinder.

Analyses of the results performed on the baseline cylinder showed that the above mentioned temperature irregularities occurred only in a limited - upper part of the cylinder wall, and that maximum temperature values occurred in the vicinity of the exhaust channel. Solution of the problem should therefore include intensive local and controlled heat transfer. A horizontal curved channel fed with lubrication oil and introduced into the critical upper part of the air cooled cylinder wall was used to save the problem. Dimensions, position and intensive heat transfer in the laminar oil flow in the curved channel helped to reduce maximum cylinder temperatures.

Circumferential wall temperature assymetry was reduced by unequal and controlled heat transfer within two channel branches. Longitudinal temperature gradients were simultaneously reduced to one third of their initial value. The oil cooling component is self-controlled: increase of the engine load is mainly compensated by the simultaneous increase of the heat transfer coefficient and the wall temperatures remain more or less unchanged. Cooling oil flow required for optimum combined cooling presents no special problem; oil pump capacity is slightly increased. The heat released into the cooling oil requires a bigger capacity oil-cooler. However, the resulting increase of the oil temperature in the oil pan does not exceed a few degrees centigrade.

cooling		engine load [%]	ΔT/Δl section '3'	[°C/mm] section '8'
○ ——	air	100	0,82	0,42
□ ——	air+oil	100	0,27	0,24
△ ——	water	100	0,59	0,19

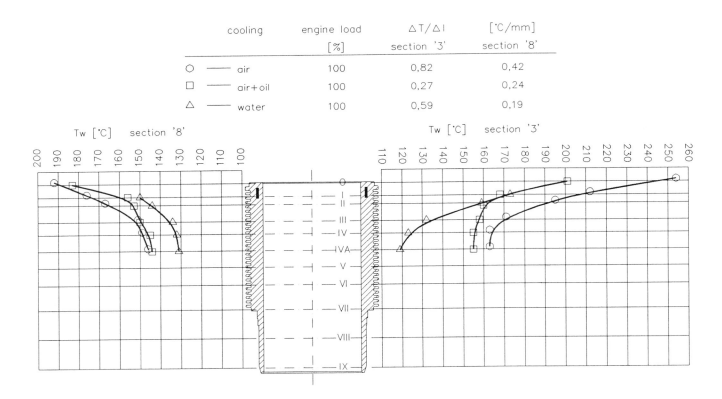

FIG.8 TYPICAL LONGITUDINAL TEMPERATURE DISTRIBUTION FOR THE AIR COOLED, COMBINED AND WATER COOLED CYLINDER AT THE SAME ENGINE RUNNING CONDITIONS

CONCLUSIONS

A new combined air-oil cooling system applied on the prototype TAM TC & iC Diesel engine cylinders is presented in the paper and its main features are:

- The intensity of the additional oil cooling follows local requirements for cooling.
- Construction of the channel and the control of the cooling oil flow awithin the channel branches is very simple.
- Optimisation of the combined cooling system results in the decrease of maximum rated and local cylinder wall temperatures by almost 40°C, critical longitudinal cylinder wall gradients are reduced by more than three times and maximum circumferential wall temperature differences by nearly 40°C.
- New temperature distribution of the combined cooled cylinder makes possible furthermore an additional and safe increase of the engine specific power by more than 10%; thermal level of the newly designed cylinder is in the same time identical to the level of the baseline air cooled cylinder, but at the engine load, that is as low as 67% of the rated value.
- Temperature distribution in the upper part of the combined cooled cylinder wall is very similar to that of the typical water cooled one.
- Quantity of the cooling air is less important by the combined cooling; thr capacity of the cooling air fan can therefore be reduced together with the required driving power.

- Reduction and redesign of the cylinder cooling fins makes possible simpler, more rigid construction and an even larger cylinder bore.
- No special additional investments are required for the production of the combined cooled engines. Bigger and competitive rated specific power and more reliable operation of the engine can be expected in the same time. Low specific fuel consumption, good ecological picture (complying with EURO1 requirements) and all other specific advantages of the air cooled engines give favourable prospects for the air cooled engines in the future.

ACKNOWLEDGEMENTS

The authors wish to record their appreciation of the TAM R&D Institute, which permitted the publishing of this paper. Moreover, the valuable financial support of the Ministry for Science and Technology of the Republic of Slovenia is greatefully acknowledged.

REFERENCES

Damjanić,F.B., and Owen,D.R.J.,1982, "Practical Considerations for Thermal Transient Finite Element Analysis Using Isoparametric Elements", *Nuclear Enginnering and Design*, Vol.69, pp.109-126

Damjanić, F.B., 1991, "THENOL" Package, Version 2.01, Manual for Users, FAGG - University of Ljubljana, Slovenija,

Owen,D.R.J.,and Damjanić,F.B., 1983, "Reduced Numerical-Integration in Thermal Transient Finite Element Analysis", *Computers and Structures*, Vol.17, pp.261-276

Šelih, J., 1990, "Numerical Methods for Engineers - Thermal Analysis of the Air cooled Engine Cylinder", Magister Work No. 70, FAGG-University of Ljubljana, Slovenija

Trenc, F., 1992, "Analysis of the Temperature Distribution in an Air Cooled Diesel Engine", Dr.Thesis No.137/D, Faculty of Mechanical Engineering - University of Ljubljana, Slovenia

NEW BEARING TYPES FOR DIESEL ENGINES:
MATERIALS, PROCESSES, APPLICATION

U. Ederer, H. Kirsch, F. Koroschetz
Miba Gleitlager AG
Laakirchen, Austria

PREFACE

This paper is an update of one originally presented at the 19th CIMAC - Congress in Florence, 1991.

It has been awarded the citation as the best Diesel engine related paper of this Congress.

The paper is presented at the 1992 I.C. Engine Fall Conference of ASME with the kind permission of CIMAC.

ABSTRACT

Only bearing materials that have an extremely fine crystalline structure and that avoid intermediate layers, enable any significant advance with respect to the fatigue stength of the layer materials and the tribological properties of the bearings. This is shown on the example of the sputtered AlSn20 overlay. Bearing types with these materials open new dimensions for the design of internal combustion engines. Test results in three engine types are presented. An application profile can be established for the most important bearing types which permits selection of the optimal bearing for every engine.

1. INTRODUCTION

Miba develops and produces engine bearings primarily for four-stroke diesel engines, and this almost exclusively for commercial vehicles and larger engines.

Demands on increased bearing performance that make further development necessary, can result from:
- changing operating conditions in existing engines (e.g. extended service intervals, heavy fuel oil operations, etc.)
- further development of existing engines
- new engine designs

2. STATE OF THE ART

2.1. General Considerations

Main and connecting rod bearings in diesel engines are subjected to high pulsating loads. The transfer of force and high surface velocity demand mutually contradictory material properties that cannot be fulfilled by a single bearing material. Modern bearings thus consist of several layers, each with a specific function [1]

In any case a steel shell is required to guarantee the press fit in the bearing housing.

The bearing materials used in bimetal bearings, such as white metal and AlSn20, are heterogeneous in their structure and, basically, consist of soft and hard components. By the soft component good operating properties under mixed friction is achieved such as wear-in performance and adaptability as well as minimal propensity to seize and low vulnerability to foreign particles. The hard component produces wear resistance and fatigue strength.

Bimetal bearings (Fig. 1) are suitable only for low loads and are mainly used as crank shaft bearings for large four-stroke engines or as cam shaft bearings in the market segments important to Miba.

In order to guarantee the fatigue strength under high loads, high strength bearing materials are necessary. Their tribological properties can, at best, be considered as emergency running capability. Thus an additional soft electroplated overlay is required. Furthermore, a nickel intermediate layer is necessary to facilitate bonding or to impede diffusion [2].

This so-called trimetal bearing (Fig. 2) continues to be the most frequently used bearing type in diesel engines.

The disadvantages of this trimetal bearing are the soft overlay's insufficient resistance to wear and the hard nickel layer, which causes a decline in the emergency-running capability of the bearing after the overlay is completely worn off [3].

There are two approaches to eliminate these drawbacks, i.e. too

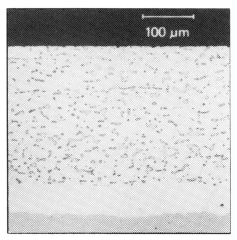

Fig. 1: Bimetal bearing Steel/AlSn20

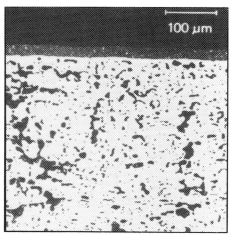

Fig. 2: Trimetal bearing
Steel/CuPb22Sn/Ni/PbSnCu

little fatigue strength of bimetal bearings and too little wear resistance of trimetal bearings, without sacrificing good tribological properties:

a) Geometric design:
   By achieving a suitable separation of hard and soft components on the bearing surface, as in the Rillenlager [3].

b) Material technology:
   By creating a refinement of the micro-structure of the bearing material, i.e., by using a fine grain texture with the distribution of hard and soft components as uniformly as possible. This improves not only the fatigue strength and wear resistance, but also the tribological properties.

2.2. Rillenlager

Ten years of successful application of the Miba-Rillenlager confirm the advantages expected from a bearing type with optimized distribution of hard and soft components on the running surface [3, 4, 5, 6, 7, 8, 9, 10, 11].

The advantages of the Rillenlager design (Fig. 3) over conventional trimetal bearings are:

- much higher wear resistance while maintaining good tribological behaviour (adaptibility, insensitivity to foreign particles)
- high resistance to corrosion - erosion in HFO-operation [2, 11]
- A maximum exposure of the Nickel dam of 5 % on the running surface over the entire bearing life. Only after the groove pattern is completely worn or flattened, the Nickel share can reach approx. 50 %. [12]

2.3. The AlZn4.5SiPb Bearing Material

For high-loaded bearings, not only the development of a wear-resistant overlay but also of an appropriate thermally stable bearing material with high fatigue strength at elevated temperatures has been necessary.

The best variant proved to be AlZn4,5SiPb.

Fig. 4 shows the microstructure of AlZn4,5SiPb and AlSn6 onto the steel.

The static strength properties e.g. yield point and tensile strength of AlZn4,5SPb at 150°C have been increased about 35 % over that of the standard alloy AlSn6, cf. Table 1.

The fatigue strength of all aluminum standard materials is documented in Fig. 5: here the alternate bending strength at ambient temperature of the steel-aluminum compound is given.

Fig. 5 also shows the penalty in fatigue strength when using a bonding foil of aluminium.

3. THE SPUTTERED OVERLAY AND NEW BEARING TYPES

3.1 Introduction.

For several further developments and new designs of diesel engines, bearings with significantly higher load limits than a oilfilm peak pressure of 400 N/mm², corresponding to approx. 60 N/mm² specific pressure, are required. At the same time, the expectations on service life have drastically increased, while reliability of operation cannot be compromised.

A major advance in both mechanical and tribological properties of the overlay is in demand. A material is required that simultaneously fulfills the tasks of the soft electroplated layer and the harder cast or roll bonded layer.

A coating process had to be selected that was capable of creating finest soft embeddings in a harder, very fine-grain matrix.

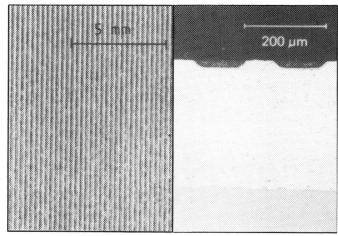

Fig. 3: Miba-Rillenlager Structure of running surface
a) View of the surface
b) Cross section perpendicular to running direction

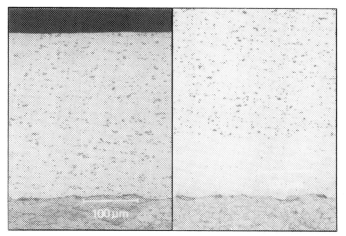

AlZn4SiPb/steel   AlSn6/Al/steel

Fig. 4: Cross section, microstructure

These crystallographic properties can be achieved with a PVD process (physical vapor deposition).

Of all known PVD processes, the sputtering process is optimal. Most materials can be deposited by this method and excellent bonding is achievable.

In order to keep the development period short, AlSn20 was selected as the material for the first generation of sputtered bearings because its good tribological properties were already known from bimetal bearing applications.

3.2 The Principle of Sputtering.

When high-energy particles, e.g., inert gas ions, impact onto a

Bearing material	Strength properties of the materials in a condition equal to the final bearing.									
	Rp 0,2 [N/mm²]		Rm [N/mm²]		A₅ [%]		E₀-Modulus [N/mm²] . 10³		Hardness [HB]	
	20°C	150°C	20°C	150°C	20°C	150°C	20°C	150°C	20°C	150°C
CuPb22Sn cast	90 bis 120	n.b.	210 bis 250	n.b.	6 bis 12	n.b.	80 bis 85	n.b.	45 bis 75	40 bis 65
AlSn6CuNi rolled	55 bis 65	40 bis 50	135 bis 145	95 bis 105	25 bis 30	40 bis 60	68 bis 73	61 bis 70	35 bis 45	25 bis 30
AlZn4SiPb rolled	95 bis 115	70 bis 100	175 bis 195	125 bis 145	18 bis 26	25 bis 32	72 bis 76	64 bis 74	45 bis 60	35 bis 40

Table 1

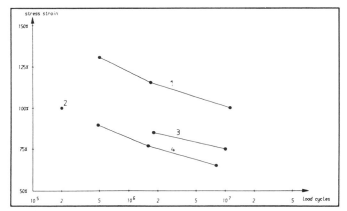

Fig. 5. Fatigue strength obtained on an alternate bending test rig
(1) AlZn4SiPb/Al/steel   (2) AlZn4SiPb/steel,
(3) AlSn6/Al/steel,   (4) AlSn20/Al/steel.

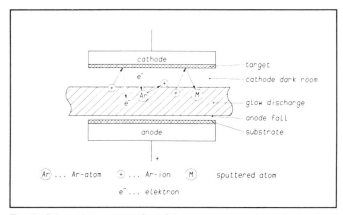

Fig. 6: Schematic representation of the sputter process in its diode configuration

solid surface, the impulse transfer causes the emission of atoms from the target material. This sputtering serves as the basis of a coating process that is carried out in a vacuum.

In the most common sputtering configuration, the diode configuration (Fig. 6), a glow discharge - maintained between the coating material (target) and the material to be coated (substrate) - serves as the ion source. The ions that are formed in the plasma of the discharge by collision ionization are accelerated in the electrical field toward the cathode (target) and cause the emission of atoms and electrons from the target.

While the electrons take on energy in the electrical field and thus contribute to the maintainance of the glow discharge, the freed atoms, unaffected by the electrical field, pass through the discharge area and condense on the surrounding surfaces.

The biggest drawback of this "diode" sputtering, its very low rate of deposition, was only overcome recently with the development of "magnetron" sputtering. By means of a special magnetic field, this process keeps the electrons emitted from the target on a closed path in the vicinity of the cathode. Since the electrons can leave this path only through impact, their energy is almost completely applied to the ionization process. This increases the number of ions that impact on the target and thus also increases the deposition rate.

Modern magnetron sputtering processes enable coating rates of 0,5 to 2,5 μm/min, depending on the target material used.

3.3 Development of the Process.

A cylindrically symmetrical configuration with an internal rod-shaped sputter source is ideal for the uniform coating of the running surface of bearings. This is the only way to achieve a precisely dimensioned coating even with thick layers as well as an optimal, uniform removal of material from the target.

Units with cylindrical sputter systems were not and are not offered by the producers of vacuum units.

In an exemplary cooperative project between Miba and the Technical University of Vienna *), a pilot unit with a rodshaped magnetron cathode tailored to the coating of bearings was designed and constructed from 1984 to 1985.

Experience gained on the pilot unit was invested in the development and construction of an initial series-production unit.

3.4 How the Process Works.

The vacuum processes are run automatically with the specified process parameters.

An absolute prerequisite for optimal bonding of the coating is the complete absence of contaminants of the surface to be coated. The sputtering process affords an elegant and relibale method for removal of gas atoms adsorbed to the surface and of the oxide layer: sputter cleaning or ion etching.

During sputter cleaning the bearing shell is poled as cathode and is thus bombarded with ions. Gas and oxide particles are sputtered off. The gases are sucked off by the vaccum pump and the oxide particles deposit on a cylindrical device that was especially developed by Miba [13].

Throughout the entire process, sputter cleaning and the succeeding coating, the process parameters are controlled, monitored and protocolled. This assures reproduceable quality of the bonding and the layer.

In addition, the actual layer thickness is computed from the electrical discharge data, so that precise coating within narrow tolerances is possible. Fig. 7 shows a view of the sputter chamber during the sputtering process with the cylindrical magnetron cathode developed by Miba [14].

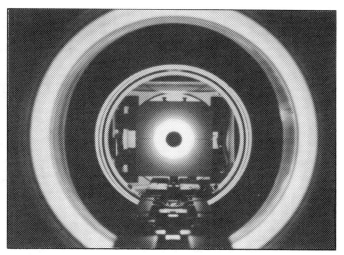

Fig. 7: View into the sputter chamber: cylindrical magnetron cathode with target (center dark), ignited plasma (bright) and the bearings to be coated in the substrate holder.

3.5 Properties of Sputtered AlSn20 Layers.

The structure of a sputter layer depends primarily on the relationship of substrate temperature to melting temperature of the coating material, and on the pressure of the sputter gas [15].

As the substrate temperature increases, the coating becomes more coarse-grained and rougher (Fig. 8a-c). Likewise, a too high content of other gases (e.g., oxygen) in the sputter gas (argon) can cause the formation of undesirable surface structures and so-called whiskers (Fig. 8d).

Just how fine the grain of a sputtered layer is compared to rolled AlSn20 can be seen in the contrasted grain structures in Fig. 9.

The AlSn20 sputter layers produced by Miba characteristically have an average grain size of the soft Sn embedments of less than 3 m and a structure of the sputter layer matrix that is basically perpendicular to the bearing surface [16].

This directed structure increases the compressive strength of the material and reduces the possible size of abrasive particles and thus the rate of wear. The uniform distribution of the fine Sn embedments in the fine-grain AlCu matrix causes a low notch factor and a high fatigue strength.

*) Institute for Applied and Technical Physics, Department for the Physics of Thin Coatings

Fig. 8: SEM, Surface, Structure of the AlSn20
sputter layer
    a)   substrate temperature too low:
        crystal too fine, too brittle
    b)   desired temperature: fine crystal
    c)   substrate temperature too high:
        crystals too coarse
    d)   oxygen content too high at desired
        temperature: formation of whiskers

(a)

(b)

(c)

(d)

Fig. 9: Cross section; Grain boundary etched
Comparison of the grain size and the
size of tin embedments in rolled
AlSn20 (a) and sputtered AlSn20 (b)

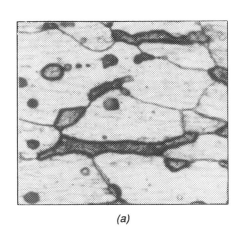

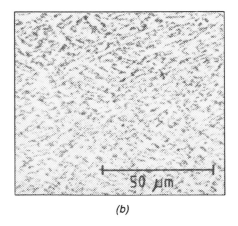

(a)

(b)

3.6 Sputtered Bearing Types.

Miba currently produces three types of bearings with sputtered overlay:

- Steel/AlZn4,5/AlSn20 (sputtered)

This type represents an ideal material combination because no brittle intermetallic diffusion phases are created even at high engine operating temperatures. The hard diffusion barrier is not necessary.

The homogenous interface between the bearing material (AlZn4,5) and the sputter layer (AlSn20) is shown in Fig. 10.

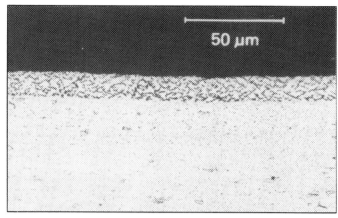

Fig. 10: Cross section, NaOH etched, Layering: AlSn20 sputtered onto AlZn4,5

- Steel/CuPb22Sn/Ni (sputtered) / AlSn20 (sputtered)

Contrary to the above type, the deposition of an AlSn20 sputter layer to lead bronze requires the application of a sputtered Ni diffusion barrier. This structure is shown in Fig. 11.

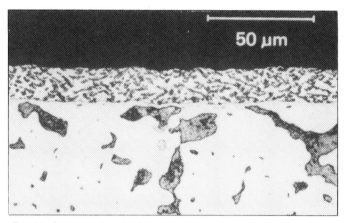

Fig. 11: Cross section, NaOH etched, Layering: AlSn20/Ni sputtered onto CuPb22Sn

- Steel/AlSn20 (sputtered)

An AlSn20 sputter layer approximately 30μm thick, deposited directly onto the steel support shell, as shown in Fig. 12, guarantees the maximum fatigue strength and invulnerability to extreme edge pressures.

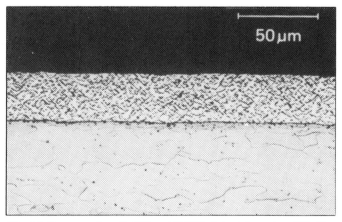

Fig. 12: Cross section, NaOH and HNO3 etched, Layering: AlSn20 sputtered onto the steel

These three types of bearings combine the best of tribological properties with the highest wear resistance and fatigue strength. These excellent properties are confirmed by test results both on the bearing test rig [17] and, more important, in engines.

4. RESULTS OF ENGINE TESTS
4.1. Medium-Speed Engine

Bearing type:	Bimetal-bearing	Trimetal-bearing	Rillenlager
Engine data	P = 200 kW/Cyl. pme = 17,93 bar n = 1000 min^{-1} Oil = SAE 40	P = 300 kW/Cyl. pme = 19,7 bar n = 600 min^{-1} Oil = SAE 40	P = 300 kW/Cyl. pme = 19,7 bar n = 600 min^{-1} Oil = SAE 40
Bearing data	Oiltemp. = 80°C p = 30,5 N/mm² $\bar{p}_{max}$ = 111,1 N/mm² h_{omin} = 4,8μm	Oiltemp. = 95 °C p = 32,5 N/mm² $\bar{p}_{max}$ = 187 N/mm² h_{omin} = 3,7μm	Oiltemp. = 95 °C p = 32,5 N/mm² $\bar{p}_{max}$ = 187 N/mm² h_{omin} = 3,7μm
Running time	10.000 h heavy fuel operation	10.000 h heavy fuel operation	12.000 h heavy fuel operation
Condition of the running surface:			
	The primary load zone shows impregnated oil coke particles and slight grooves. No wear is visible. Determination of the amount of wear is possible only by measuring the difference in wall thickness.	The electroplated overlay (original thickness 30μm) is eroded to the nickel dam in large areas in the primary load zones. The transitions from the worn zones to the overlay are visible as dark corroded regions.	The width of the lands between grooves has not changed in the primary load zones. There is distinct wear at the center of the bearing. The overlay in the grooves is worn app. 5μm, the lands not more than 2 μm. Oil coke particles are embedded in the overlay. Small grooves from foreign particles are visible.
Evaluation of test results:			
Still usable	Yes	No	Yes
Estimation of remaining service life	Impossible	Failure risk too high; replace	Only one fourth of service life expired
Failure of bearing expecte due to	Fatigue, seizure	Seizure	Grooves are smoothed, seizure can occur

Cross section of the primary load zone

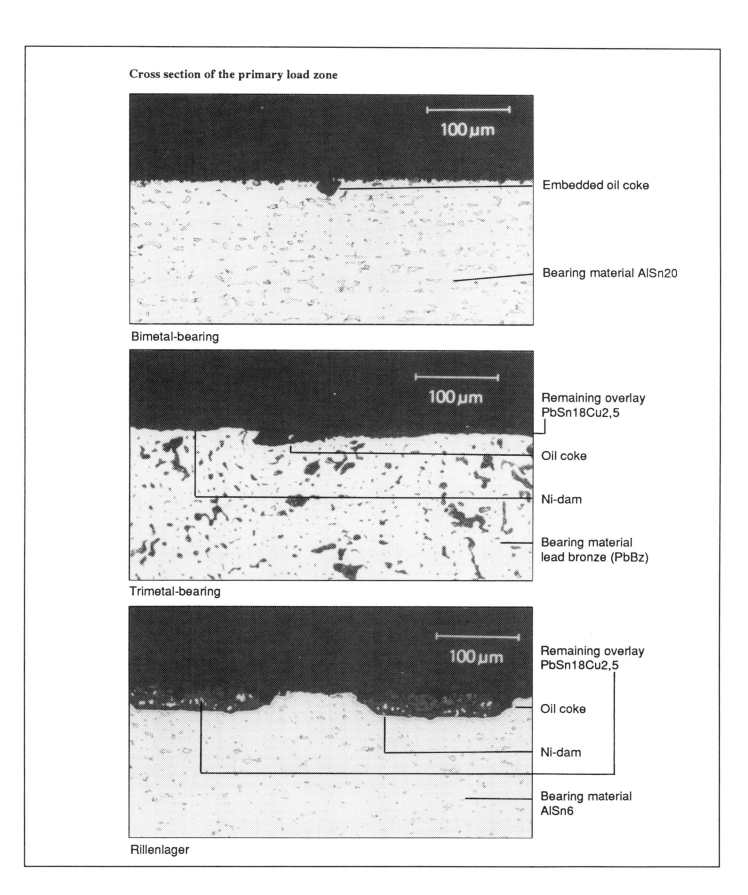

Embedded oil coke

Bearing material AlSn20

Bimetal-bearing

Remaining overlay PbSn18Cu2,5

Oil coke

Ni-dam

Bearing material lead bronze (PbBz)

Trimetal-bearing

Remaining overlay PbSn18Cu2,5

Oil coke

Ni-dam

Bearing material AlSn6

Rillenlager

4.2. Large High-Speed Engine

Bearing type:	Trimetal-bearing	Rillenlager	Sputtered-bearing
Engine data	P = 120 kW/cyl. pme = 17,3 bar n = 2100 min^{-1} Oil = SAE 15W40 pz = 135 bar	P = 120 kW/cyl. pme = 19,3 bar n = 2100 min^{-1} Oil = SAE 15W40 pz = 135 bar	P = 120 kW/cyl. pme = 19,3 bar n = 2100 min^{-1} Oil = SAE 15W40 pz = 135 bar
Bearing data	Oiltemp. = 130 °C p = 50 N/mm^2 $\bar{p}_{max}$ = 300 N/mm^2 h_{omin} = 1,3 μm	Oiltemp. = 130 °C p = 50 N/mm^2 $\bar{p}_{max}$ = 300 N/mm^2 h_{omin} = 1,3 μm	Oiltemp. = 130 °C p = 50 N/mm^2 $\bar{p}_{max}$ = 300 N/mm^2 h_{omin} = 1,3 μm
Running time	500 h	10.000 h	10.000 h
Condition of the running surface:			
Evaluation of test results:	The electroplated overlay is locally worn to the nickel layer in the primary load zones. Regions with remaining overlay display fatigue fractures due to overload.	The overlay in the grooves is almost completely worn. The lands between grooves show wear of about 3 μm.	The sputtered layer shows hardly any wear. Wall thickness measurements revealed a total wear of 3 μm which results from evening out the surface roughness.
Still usable	No	Yes	Yes
Estimation of remaining service life	The bearing has no conformability left; reuse is not possible.	The bearing has sufficient reserves to continue to operate. Another 5.000 hours should be possible without problems.	The expected service life cannot yet be quantified from the state of wear or from experience to date. At least another 20.000 hours of service life is to be expected.
Failure of bearing expected due to	Seizure	Seizure	Fatigue

Cross section of the primary load zone

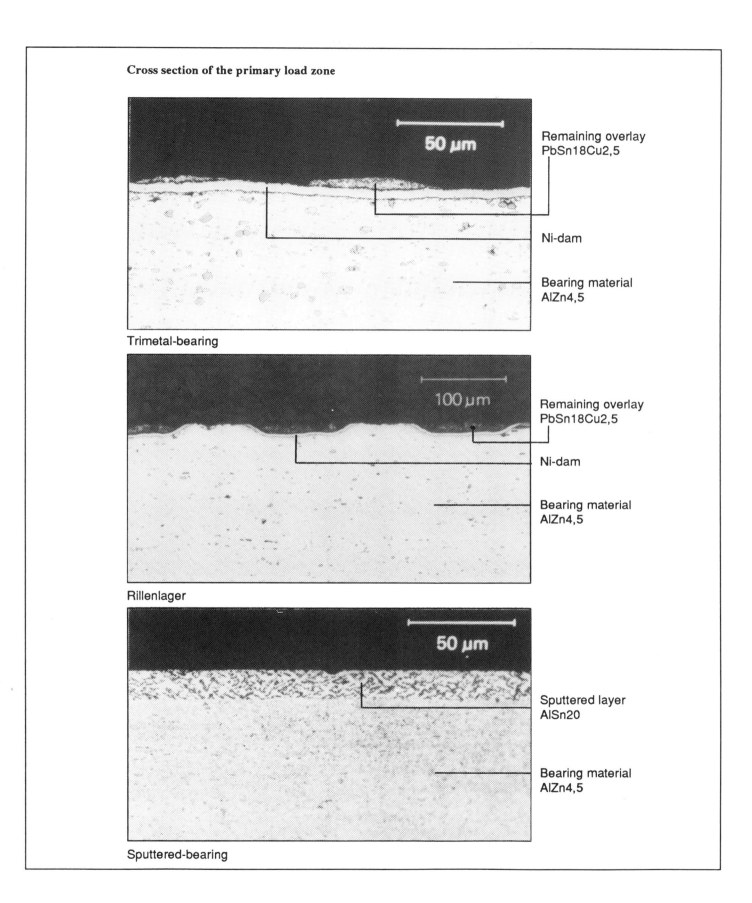

50 µm

Remaining overlay
PbSn18Cu2,5

Ni-dam

Bearing material
AlZn4,5

Trimetal-bearing

100 µm

Remaining overlay
PbSn18Cu2,5

Ni-dam

Bearing material
AlZn4,5

Rillenlager

50 µm

Sputtered layer
AlSn20

Bearing material
AlZn4,5

Sputtered-bearing

43

4.3. Engine for Commercial Vehicles

Bearing type:	Trimetal-bearing	Rillenlager	Sputtered-bearing
Engine data	P = 40 kW/cyl. pme = 15,2 bar n = 2100 min^{-1} Oil = SAE 15W40	P = 40 kW/cyl. pme = 15,2 bar n = 2100 min^{-1} Oil = SAE 15W40	P = 40 kW/cyl. pme = 15,2 bar n = 2100 min^{-1} Oil = SAE 15W40
Bearing data	Oiltemp. = 130 °C p = 61 N/mm^2 $\bar{p}_{max}$ = 380 N/mm^2 h_{omin} = 0,8 μm	Oiltemp. = 130 °C p = 61 N/mm^2 $\bar{p}_{max}$ = 380 N/mm^2 h_{omin} = 0,8 μm	Oiltemp. = 130 °C p = 61 N/mm^2 $\bar{p}_{max}$ = 380 N/mm^2 h_{omin} = 0,8 μm
Running time	500.000 km	500.000 km	500.000 km
Condition of the running surface:			
Evaluation of test results:	The electroplated overlay is worn to the nickel layer.	In the primary load zones the overlay has in part been washed out of the grooves so that only traces there of remain. The lands are worn down to 8 μm, which is half of the nominal groove depth.	The sputtered bearings show traces of conforming, but no actual wear.
Still usable	Not useable due to lack of failure reserve.	The bearings have sufficient wear reserves and conformability; further use is thus possible.	This bearing can be reinstalled without any reservations.
Estimation of remaining service life	Risk of failure too high	500.000 km	Estimation not possible with current test data.
Failure of bearing expected due to	Seizure	Seizure	Estimation not yet possible.

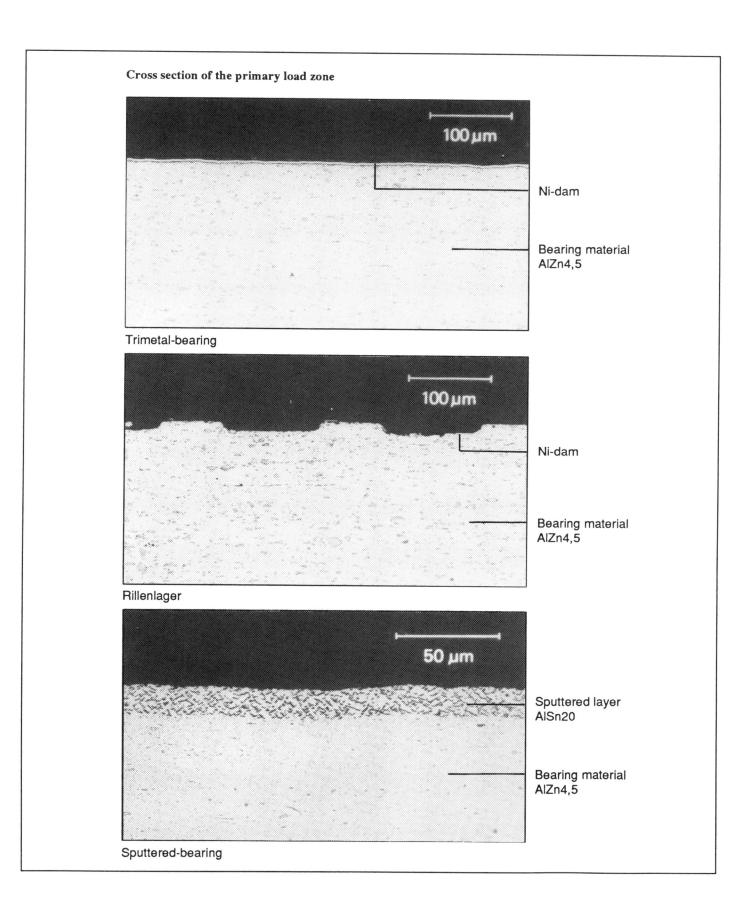

Cross section of the primary load zone

100 µm

Ni-dam

Bearing material
AlZn4,5

Trimetal-bearing

100 µm

Ni-dam

Bearing material
AlZn4,5

Rillenlager

50 µm

Sputtered layer
AlSn20

Bearing material
AlZn4,5

Sputtered-bearing

5. APPLICATION PROFILE FOR MODERN BEARING TYPES IN TERMS OF OIL FILM PEAK PRESSURE AND MINIMUM OIL FILM THICKNESS

Permissible ranges for a type of bearing - as they are given in Section 3, for example - can be determined with respect to the required service life for each size of engine on the basis of numerous applications.

The following diagrams (Figs. 13 and 14) consider 5 different types of bearings. The basis for comparison is computed values for connecting rod bearings for various engines. The size of these engines ranges from commercial vehicle to medium speed four-stroke engines.

Such diagrams serve as a basis for preselection of suitable bearing types, including considerations stemming from bearing test rig data.

The values specified here are intended as guidelines and need to be individually modified when special conditions arise. The customs of engine manufacturers (tolerances of connected parts) and the servicing of the engine (e.g. care of lubricating oil) play a role.

5.1. Fatigue Strength Limits of Various Bearing Types

The computed maximum lubricating oil film peak pressure is used to determine the necessary fatigue strength of the bearing material. (Fig. 13) [17]

The load values for bearing materials that have been used for years can be viewed as very certain. The specified limit for the steel/AlSn20 sputtered bearing represents the minimum from operational experience collected to date. It is probable that this bearing type will display even higher load limits.

5.2. Wear Limits of Modern Bearing Types

In order to select the optimal bearing type, the wear resistance of the bearing must be considered in addition to its fatigue strength. This aspect incorporates both Miba's computed theoretical minimum lubricating oil film thickness (h_{omin}) and practical experience. [17] The diagram in Fig. 15 shows the permissible limits for various bearing types on the basis of the same connecting rod bearings as above.

The diagram shows that the steel/AlSn20 bimetal bearing can be employed wherever wear resistance is called for, but the demands for fatigue strength are not too high.

It can be clearly seen that the Rillenlager is superior to the trimetal bearing in every area of application.

The sputtered bearing has the highest wear resistance, although the specific limits have not yet been determined. An evaluation of sputtered bearings over 260 mm diameter has not yet been possible because Miba has not been producing bearings in this size range. The graph of the wear resistance is thus extrapolated with a dotted line.

6. CONCLUSIONS

The development of the highly wear resistant sputtered overlay result in advanced bearing types. Both the fatigue strength of the lining and the tribological properties of the running surface are significantly improved.

The sputtered bearing in its presented types permits increases in specific power or in time between service intervals. We are convinced, nevertheless, that this bearing type has a high potential for further development.

The example of the Miba sputter project demonstratively showed that the intensive cooperation of science, producer and user are necessary in order to succeed. Decades of excellent information feedback make it possible for us to set clear application guidelines. Miba's evaluation of the results of theoretical computations starting in the design stage has thus gained a high reliability.

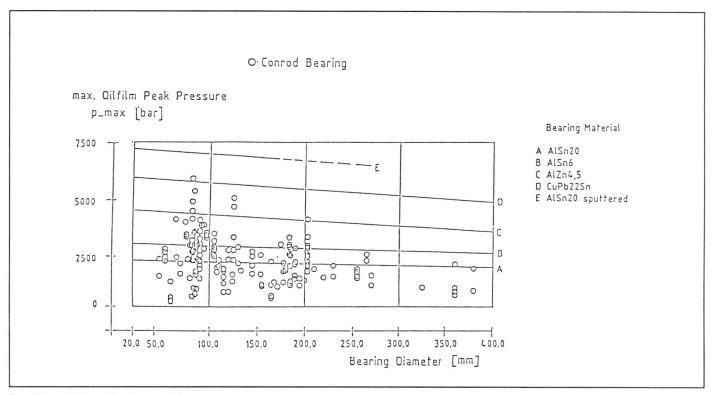

Fig. 13:Load limits of bearing materials;
Maximum permissible lubricating oil film pressure over bearing diameter.

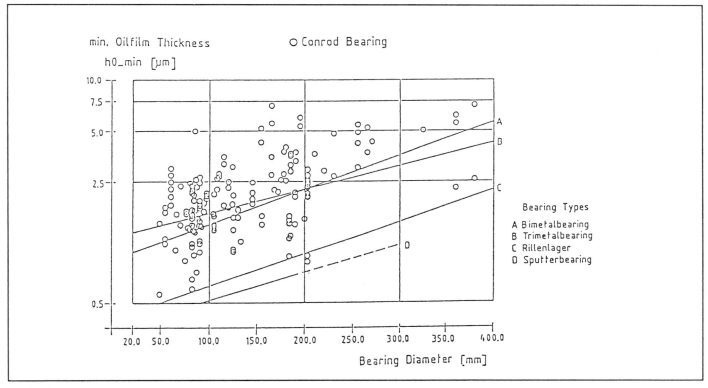

Fig. 14:Limits for wear;
Minimum permissible lubricating oil film thickness over bearing diameter.

REFERENCES:

[1] Koroschetz, F.: Bearing Materials in Development, Production and Application.
Miba-Symposium 1982. (Page 2)

[2] Koroschetz, F., Hocke, E. und Gasserbauer, M.: Diffusionsvorgänge an galvanisch abgeschiedenen PbSnCu-Gleitlagerlaufschichten. Microchemica Acta (Wien) 1981, Suppl. 9, S. 139-152. (Page 2)

[3] Ederer, U. G.: New Bearing Designs for Higher Performance Requirements.
Miba-Symposium 1982. (Page 3)

[4] Ederer, U. G.: Lager für hohe Belastungen in Zweitakt- und Viertakt- Dieselmotoren.
MTZ 44 (1983) 11, S. 443-450. (Page 3)

[5] Grobuschek, F.: New Bearing Designs.
SAE-Paper 845091, XX. FISITA CONGRESS 9.-11. MAI 1984, WIEN. (Page 3).

[6] Ederer, U. G.: Bearings for Higher Performance Requirements.
ASME-Paper 84-DGP-5, 1984. (Page 3)

[7] Ederer, U. G.: Design Criteria and Experience with Crankshaft Bearings in Medium Speed Engines.
SAE-Paper 851196, Government/Industry Meeting & Exposition Washington, D.C., May 20-23, 1985. (Page 3)

[8] Ederer, U. G.: Neue Lagerbauarten für gesteigerte Anforderungen.
Tribologie und Schmierungstechnik 33 (1986) 5, S. 264-271. (Page 3).

[9] Ehrentraut, O.: Hochbelastbares Gleitlager. Österr. Patent Nr. 369149, ert. 1982, Europapatent Nr. 0057808, ert. 1983, US-Patent Nr. 4 440 099, ert. 1983. (Page 3)

[10] Grobuschek, F.: Optimale Gleitlagerungen.
Tribologie und Schmierungstechnik 32 (1985) 4, S. 188-191. (Page 3)

[11] Grobuschek, F.: Einsatz von Rillenlagern in Viertakt-Dieselmotoren im Schwerölbetrieb.
MTZ 48 (1987) 4, S. 141-145. (Page 3)

[12] Miba (anonym): Miba Replacement Criteria for the Rillenlager. Miba 1989, Broschüre. (Page 3)

[13] Koroschetz, F., Gärtner, W., Wagendristel, A., Bangert, H.: Vorrichtung zur Behandlung von Innenflächen mit Ionenbeschuß.
Österr. Pat. No. 387 989, Pat.: Feb. 14, 1989, Europ. Appl. No. 0310 590, Appl.: July 20, 1988. (Page 8)

[14] Koroschetz, F., Gärtner, W., Wagendristel, A., Bangert, H.: Stabförmige Magnetron- bzw. Sputterkathoden-Anordnung.
Österr. Appl. No. 7A 2193/87, Appl.: Sept. 1, 1987 Europ. Appl. No. 0300 995, Appl.: July 20, 1988. (Page 8)

[15] Thornton, J.A.: ... Sructure of Thick Sputtered Coatings.
J.Vax. Sci. Technol. 11, 4 (1974). (Page 9)

[16] Koroschetz, F., Gärtner, W.: Hochbelastbares Gleitlager.
Österr. Pat. No. 389 356, Pat.: Sept. 12, 1989, Europ. Appl. No. 88 89 9188.1, Appl.: July 18, 1988. (Page 10)

[17] Aufischer, R., Ederer, U. G.: Improved Methods for the Prediction of Relibility of Engine Bearings.
CIMAC 91, Italy. (Page 11)

INTEGRATED GAS ENGINE MANAGEMENT AND CONTROL

Carl F. Vaske
Engine Controls Group
Woodward Governor Company
Fort Collins, Colorado

ABSTRACT

The evolution of microcontrollers over the past decade has provided a host of solutions to engine control and management. This paper details a new microcontroller based integrated gas engine manager that is designed for control of large gaseous fueled spark ignited engines that are employed in pipeline compression and electrical power generation applications. The major focus is on the flexibility of this system to address a wide range of engine control tasks without the complexity typically associated with digital electronic controls, that is, system configuration and program modification. A secondary yet equally important discussion on control communications is also covered.

INTRODUCTION

Historically, control systems for large gas engine applications in pipeline compression and electrical power generation have been composed of pneumatic or hydro-mechanical control devices. In some cases the control systems are a hodgepodge of analog electronic control devices, relays, switches, etc. With the advent of the microprocessor and its offspring the microcontroller, control systems have typically been composed of a Programmable Logic Control (PLC) to handle system sequencing and monitoring, while deferring the time critical tasks to older analog or simple real time digital control devices.

Owners and operators of facilities using gaseous-fueled spark-ignited (GFSI) engines have been encouraged to make their plants more environmentally clean and economically efficient due to the enactment of recent legislation. In particular, the 1990 Clean Air Act addressed environmental issues, while legislation passed during the 1970s and 1980s deregulated gas pipelines (spurring competition) and opened accessibility to electric utilities (leading to a proliferation of electrical co-generation facilities). To improve their plants means updating the existing mechanical control systems and/or early generation electronic controls with newer digital electronic controls providing a higher degree of communication, plant automation and plant protection. Designing and implementing a control system with these features is a task for which most plant superintendents have neither time nor tools. The scope of this paper is to present a new electronic control device designed to meet this challenge: IGEM•1 (Integrated Gas Engine Manager). A discussion on the necessity for integrated control functions is given first. The next few sections cover the IGEM•1 hardware, followed by a discussion on the programming capabilities of an IGEM•1. Next, networking IGEM•1 for system management is covered. A look at what electronic control systems of the near future may be like is offered, and lastly, a summary of experiences gained during the development of IGEM•1 is given.

SYSTEM INTEGRATION

The logical method of controlling a complex system is to break it down into small manageable subsystems. So it is with control systems used in gas transmission or electrical power generation. A basic control system often consists of a speed control and actuator for fuel metering, with separate autonomous ignition control. In some cases the system has a PLC for start/stop sequencing and/or system monitoring. For power generation applications a sequencer/load control is usually installed. For gas transmission applications a process control may be employed. A degree of integration among various functions is required no matter how many subsystems there may be. System complexity is dependent upon the amount of integration that must occur between the subsystems.

As the power of microprocessors increased, the trend in control systems design has been to integrate more processes into a single package. PLCs are a prime example of this migration.

However, there are some drawbacks to this type of design. Among them are the number of I/O terminations that must be made at a single point. To address the termination problem, these controls are typically rack mounted chassis designs, using multiple I/O boards and a mother-board connecting I/O boards to external terminations. This means that users of a basic control system, one with a minimum number of I/O boards, absorb extra costs that large control systems require. Another drawback of rack mounted controls is that they must usually be mounted in control room environments or mounted in custom-built environmentally sealed enclosures.

With the large amount of I/O processed by rack mounted controls, designers end up coding control algorithms into one massive program. Instead of having small programs aimed only at subsystems, this singular large program is directed at controlling the entire system. In effect, this deprives the designer of the more systematic divide and conquer methodology of designing the complete control system.

The concept of IGEM™ is to provide a family of products that allow the control system designer to select only control functions he deems necessary for his application, without paying a penalty for system features not required. Table 1 lists functions the IGEM product line could be capable of controlling.

TABLE 1 IGEM CONTROL FUNCTIONS

ENGINE CONTROL	COMPRESSOR CONTROL
SPEED CONTROL	LOAD/TORQUE CONTROL
AIR/FUEL RATIO CONTROL	POCKET CONTROL
SPARK IGNITION TIMING	PROCESS CONTROL
DETONATION DETECTION	
SYSTEM LEVEL CONTROL	**GENERATOR CONTROL**
MONITOR AND SAFETY	SYNCHRONIZER
ANNUNCIATOR PANEL	VOLTAGE REGULATOR
OPERATOR CONTROL PANEL	KVAR/POWER FACTOR CONTROL
LOCAL OPERATING NETWORK (LON)	IMPORT/EXPORT CONTROL
SEQUENCER (START/STOP)	LOAD CONTROL
SEQUENCER (PLANT MANAGER)	ISOCHRONOUS LOAD SHARING
SERIAL I/O INTERFACE	REAL POWER SENSING

IGEM products also provide a means to enhance system capabilities by utilizing Echelon LON™ (Local Operating Network). LON allows individual control modules of the IGEM product line to be integrated into a coherent control system. LON's main advantage is resource sharing that is not possible with analog controls or non-network digital controls. For example, since speed is sensed by the module containing the speed control function, it can provide speed information to other modules within the system. The difference between the LON output and an analog signal, perhaps a 4-20 mA tachometer signal, is that the LON can send more information than just speed; it might also provide rate of speed change or switch point information.

LON can broadcast information to all network modules. It is not limited by drive capacity the way an analog signal is, and it will be more accurate due to its digital nature. Thus, for the speed input example, the number of speed sensors on an engine can be reduced by sharing speed information among system modules. In addition, speed input processing circuitry is no longer required on every module needing speed information.

There are limitations for network controls. The main limitation is in data throughput. If a network had an infinite bandwidth, we could send and receive information between modules without worrying about data accuracy or timeliness. Because networks do have a finite bandwidth we must concern ourselves with such things as transmission data rates and message collision detection. We also need to be certain that data can reach its destination in a deterministic manner. The LON used on the IGEM product line automatically provides collision detection and avoidance, and it also has a priority message handling scheme to make the network deterministic. Therefore, to prevent LON system overload, we only need to limit the amount of data sent over the network.

The IGEM product line integrates control functions requiring real-time interaction into multifunction control modules as a means of reducing network traffic. The first module of the IGEM product line integrates three of the engine control functions listed in Table 1 (speed control, air/fuel ratio control, and ignition timing control) into a single hardware platform: IGEM•1. Also integrated into IGEM•1 are two of the system level control functions: the LON function and the serial interface function.

Combining these particular functions into a single control module is understandable, since all of these functions must interact with one another in a timely fashion to sufficiently control engine speed while simultaneously optimizing either exhaust emissions or fuel economy. This does not preclude integrating other functions within an IGEM module or eliminating some of those that are presently included. Indeed, if detonation detection is required for the control system, it may be necessary to combine it with the speed control function. There is no absolute fine line as to which functions must be integrated with one another. However, by separating control functions for a GFSI engine application into smaller subsets, the programming of individual modules can be reduced.

In many instances the program for one application can be ported to another system and then enhanced by adding extra

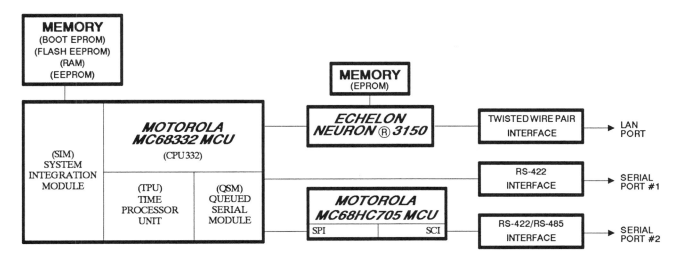

FIG. 1 IGEM CPU BOARD BLOCK DIAGRAM

control functions, for example, adding process control or KW load control to IGEM•1. Likewise, the individual engine control functions handled by IGEM•1 can work independently if an application has no requirement of the other functions. The key to system integration lies in providing an economic and flexible method of linking subsystems together. The IGEM product line accomplishes this by its unique hardware and software programming capabilities.

IGEM•1 CPU HARDWARE

The IGEM•1 CPU board is designed around one of the newest microcontrollers in the industry, Motorola's MC68332 (Fig. 1). The MC68332 is a 32 bit microcontroller designed specifically for engine control. The MC68332 MCU is based on the MC68020 microprocessor; utilizing most of the MC68020 instruction set, plus a few new instructions. The MC68332 has three intelligent subsystems on-chip that help off-load processing activity from the central CPU. This in turn enhances the computational capacity of the MC68332.

The first subsystem of the MC68332 is the Time Processor Unit (TPU). The TPU has 16 orthogonal I/O channels that can perform timing algorithms that are embedded in on-chip ROM. If the library of timing functions provided by the TPU is insufficient, the programmer can execute his own microcode out of 2 Kbytes of on-chip RAM. IGEM•1 uses two TPU channels for sensing engine speed by utilizing a time capture and compare algorithm provided by the MC68332. IGEM•1 also utilizes several TPU channels for Pulse Width Modulated (PWM) outputs. These PWM outputs are then converted to either 0-200 mA or 4-20 mA analog output signals.

The second subsystem of the MC68332 is the System Integration Module (SIM). The SIM provides an interface to external memory or other peripheral devices, while eliminating many glue logic devices typically required for microcontroller system designs. For IGEM•1 the SIM subsystem is used to

communicate to external memory and the LON processor chip. The SIM also provides accessibility to a floating point co-processor. IGEM•1's CPU board has a socket for adding a floating point co-processor if needed for future applications.

The third MC68332 subsystem is a Queued Serial Module (QSM) that provides two serial communications ports. The first QSM port is a Serial Communications Interface (SCI) that can be used for RS-232, or RS-422/485 standard communication. For IGEM•1 the SCI port is used for RS-422 communication. During the control's power-up stage, this port is checked for a connection to a personal computer (PC). If the connection exists, then IGEM•1 enters an application program down-load mode. Otherwise, this port is set up to communicate with a hand-held terminal device used for tuning and monitoring. The second QSM port is a queued Serial Peripheral Interface (SPI) port which is connected to several peripheral devices. The queue provides a bank of memory registers used for storing commands sent and data received from peripheral devices. IGEM•1 uses the SPI to interface with all Analog-to-Digital Converter (ADC) chips located on the I/O boards. It also serves as an interface to a Motorola MC68HC705 8-bit MCU that in turn is used for watchdog monitoring and RS-422/485 communications.

The memory for IGEM•1 is composed of the following: 1) Boot EPROM (64 Kbytes) that perform power-up diagnostics plus verification and/or loading of a valid application program. 2) Flash EEPROM (512 Kbytes) for application program storage. Flash memory is a key technology that provides IGEM•1 with a high degree of versatility. 3) Static RAM (128 Kbytes) for variable storage. 4) EEPROM (2 or 8 Kbytes) for storage of parameters to be retained during power down and power up cycles.

IGEM•1's CPU board also features Echelon's Neuron® 3150 chip that is the heart of IGEM•1's LON. The Neuron 3150 incorporates three 8-bit CPUs. Two of these CPUs handle LON communication protocols, while the third CPU is

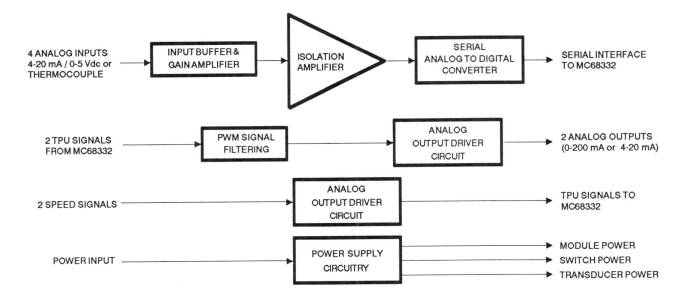

FIG. 2 IGEM I/O 1 POWER SUPPLY BOARD BLOCK DIAGRAM

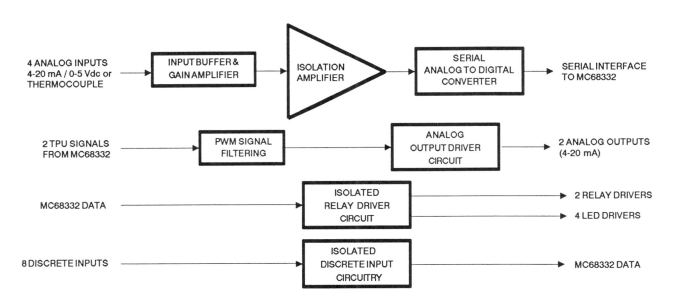

FIG. 3 IGEM I/O 2 BOARD BLOCK DIAGRAM

dedicated to user applications. The IGEM•1 LON uses a twisted wire pair transmission medium. The LON is capable of 1.25 Mbps communication over a distance of 500 meters.

IGEM•1 I/O HARDWARE

The IGEM•1 control module has three configurations depending upon the number of I/O boards installed. All three configurations have an I/O 1 power supply board (Fig. 2). This board provides the power to the CPU board and the other I/O board(s). Input power for the IGEM•1 control is 18-40 Vdc. The I/O 1 board also provides power for discrete input switches and limited transducer power.

The 'Basic' IGEM•1 configuration contains a second I/O board: I/O 2 (Fig. 3). This board is mounted atop I/O 1 and uses a VME style connector to provide board to board interconnections. Field terminations for this board come from a separate interface board mounted on a DIN rail on the side (internal) of the IGEM•1 enclosure.

I/O capabilities of the I/O 1 power supply board are as follows: 1) Two speed inputs capable of sensing either proximity probe signals or magnetic pickup signals. 2) Two analog outputs that can be configured for either 0-200 mA or 4-20 mA. 3) Two analog inputs that can be configured to sense either 4-20 mA, 0-5 Vdc, or thermocouple inputs. In addition, the thermocouple inputs can provide either high or low failure modes if the input signal has an open circuit. 4) One input channel that is identical to the other analog inputs; with an additional option for configuring it as a ±2.5 Vdc input. This input can be used as a bias input if required by the application program. 5) One input channel for sensing temperature at the point where the thermocouple wire is terminated to copper wire. This channel senses an Analog Devices AD590 temperature sensor signal and is used for cold junction compensation of the thermocouple inputs.

I/O 2 has the following I/O capability: 1) Eight optically isolated discrete inputs. 2) Two optically isolated relay driver channels. 3) Two 4-20 mA analog outputs. 4) Four analog inputs with the same capabilities as I/O 1 analog inputs.

The 'Enhanced' and 'Maximum' configurations of IGEM•1 are expansions of the 'Basic' control. 'Enhanced' controls contains one additional I/O board. The additional board, I/O 3, has the same capacity as the I/O 2 board. 'Maximum' configurations have 2 more I/O boards than the 'Basic' control.

These Enhanced and Maximum versions of the IGEM•1 are used on applications that require more I/O, for example, twin-turbocharged engines. They can also be used in applications that require more control functions than presently handled by the Basic IGEM•1.

IGEM•1 HARDWARE

The IGEM•1 CPU, I/O and field wiring Interface boards are enclosed in a NEMA 4 chassis (Fig. 4). IGEM•1 controls are designed to meet UL and CSA requirements for Class 1 Division 2 Group D environments.

All wiring to the unit enters through a gland plate on the chassis bottom. The gland plate and all connections inside IGEM•1, quick release Phoenix connectors and a DB-9 pin connector, allow for rapid replacement if required.

The CPU board is mounted to the front cover of the enclosure. It provides LED indications for power and control system status that can be viewed through the cover. The I/O boards are mounted in the body of the enclosure and are stacked on top of each other. All the boards are covered to limit direct access to electrical components and thus reduce failures that may be caused by electrostatic discharge (ESD).

SYSTEM PROGRAMMING

Application programs for the IGEM•1 are made using Application Block Language Software (ABLS) tools. ABLS is a set of PC software tools that operate under OS/2®. These

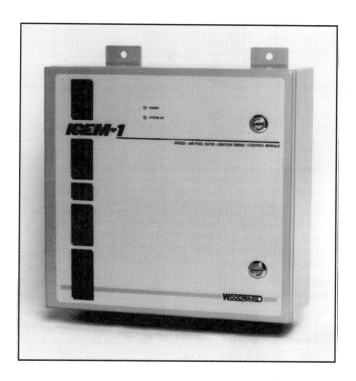

FIG. 4 IGEM•1 CONTROL MODULE

tools consist of a Menu Oriented Editor (MOE), a 'C' language code generator (Coder), a 'C' compiler, and a hex code down-load program (XFERIGEM.EXE).

MOE is an applications development platform that can be used by designers with little or no software programming experience. MOE lets the designer concentrate on control system requirements instead of programming language details, such as, data structures and syntax. MOE provides a set of building blocks (MOE blocks) that can be linked together to create a functional control application. MOE blocks have been designed and tested to handle tasks such as PID control, Boolean functions, I/O gathering, etc. Once the MOE application program has been built, it is processed by the Coder, which generates 'C' language code representative of the tasks required by each MOE block.

The 'C' program files are compiled and linked into machine language hex code which is loaded into IGEM•1's memory via the down-load program. The down-load program prompts the user to establish a communications link between a PC and the IGEM•1. Once the link is made, it asks the user for the application program's hex code filename. After the filename is entered, the process of programming the IGEM•1 begins.

IGEM•1 uses flash memory to eliminate handling of PROM devices and PROM programming with a specialized programmer station. Before the flash memory can be programmed it must be erased to prevent device degradation due to depletion mode saturation. The down-load sequence is

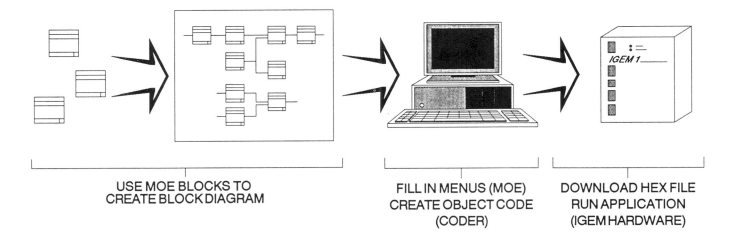

USE MOE BLOCKS TO
CREATE BLOCK DIAGRAM

FILL IN MENUS (MOE)
CREATE OBJECT CODE
(CODER)

DOWNLOAD HEX FILE
RUN APPLICATION
(IGEM HARDWARE)

FIG. 5 APPLICATION PROGRAMMING STEPS

designed so that erasure does not occur until proper communication between IGEM•1 and a PC is established and access to the specified hex file is verified. Upon completion of the application program down-load, IGEM•1 runs through its power-up diagnostics and then begins running application code. If an application program already resides in the IGEM•1 memory, the control can go through a power-down/power-up cycle and execute the application code without performing the down-load activity. Figure 5 summarizes the process required for building and programming IGEM•1.

The LON applications for IGEM are developed with Echelon's LONWORKS™, a tool set for building, testing and installing LON nodes. Each LON node uses the LONTALK™ protocol for message scheduling, LON I/O processing, network communication and network management. LON programs are written in a 'C' language derivative, Neuron®C.

LON modules may have Neuron C application programs that are transparent to the end user. When nodes are connected together to form the LON, the 'binding' process may be as simple as pressing a service button on individual nodes. This would trigger a node to broadcast a network variable list. The list tells other LON nodes what messages it is capable of receiving and sending. The other LON nodes can then use this information, as their application programs require, to logically connect themselves to the node publishing the list.

NETWORK TOPOLOGY

The LON topology is designed to use multiple domains featuring a sub-network (subnet) hierarchy, similar to having subsystems for the overall control system. If all engine level, station level, and distribution level communications were handled by a single network, it would be quickly overloaded.

Figure 6 shows a possible gas pipeline LON topology. For this topology, engine and station subnets belong to station domains. Each subnet has many nodes, perhaps control module

nodes and solenoid valve nodes, that can communicate directly to one another. For example, a compressor load control module could send a speed bias message to the speed control's node.

The station subnet and engine subnets communicate through routers. Routers provide the intelligence to send messages destined for particular addresses. A router is analogous to sending mail through the U.S. Postal Service from one company to another, versus using inter-departmental mail within a company (subnet communication). Routers are also used to communicate between domains and different types of media within a subnet. Voice mail address lists are a good example of routing within a subnet. Again, the router's sole purpose is to get messages to their proper destination.

Bridges are used for passing messages between different media. A bridge monitors all messages sent on the two media to which it is connected and forwards the messages that have destination addresses on the other side. A ferry boat dock is an analogy of a LON bridge. The dock provides the physical link between the two transportation modes, land and water. The ferry boat ticket is the traveller's pass to cross over.

If the distance between nodes, subnets, or domains exceeds the capabilities of the media, repeaters are implemented. Repeaters receive messages, boost the signal strength to their original level, then send them onward to their destination.

In some cases communication is required between a LON and networks that have different protocols, for example, tying a LON to an Ethernet network. Gateways provide a means to tie the dissimilar networks together. A gateway could be likened to a foreign language interpreter helping two parties communicate.

The LON does have some finite limitations. These include a maximum 127 nodes per subnet and 255 subnets per domain (32,385 nodes per domain total). However, since there can be multiple domains, the complete network can provide a large flexible capacity through judicious design.

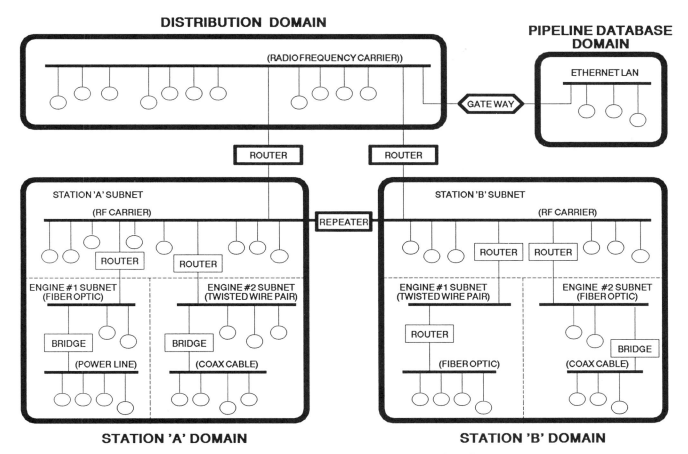

FIG. 6 GAS PIPELINE NETWORK TOPOLOGY

FUTURE TRENDS

IGEM•1 is designed to accept signals from transducers using the communications standard of the past and present, that is, 4-20 mA signals. It can also send signals to meters and actuators using 4-20 mA or 0-200 mA outputs. To produce or sense these signals costs money and printed circuit board space. Also, transporting these signals is expensive when one considers the cost of the wire used in the cabling. The IGEM product line provides a transition to the future through its LON capability. Soon, transducers, solenoids, actuators and other dedicated controls will be able to communicate with each other through LONs.

LONs will help reduce wiring costs and provide systems with greater knowledge bases than are possible today using analog signals. This new technology is destined to become the process control communications standard of the future, due to its flexibility, priority message handling scheme and low cost.

EXPERIENCES

IGEM•1 production prototypes were manufactured during the summer of 1991. One was installed on a Ruston 12 cylinder, 4-stroke, 900 RPM engine at a land-fill co-generation facility, while the second went on a Clark 6 cylinder, 2-stroke, 330 RPM engine at a pipeline compressor station.

As would be expected on a project of this nature, problems were uncovered in the field with both software and hardware. The main hardware problem dealt with a lack of adequate ground signals between the CPU board and the I/O boards. This resulted in some ground bounce that corrupted data representing the analog inputs, thus causing false sensor reading failures. These problems were remedied with a temporary field fix while the circuit boards went through a new layout.

The software bugs were twofold. One side dealt with the operating system, while the other side was related to the application program. The operating system bugs were cleaned up and a new set of ABLS tools was given to the application designer. Using the ABLS tools and the down-load capabilities of the IGEM•1, the applications designer made the changes required to fix the application program problems. All these changes were implemented in the field and the units ran successfully.

In early spring of 1992 a new productionized version of the 'Basic' IGEM•1 became available. Several changes to the hardware were implemented as a direct result of the prototype field tests. The enclosure was upgraded with a gland plate.

The I/O boards were modified to resolve the ground bounce problem. Also, the CPU board went through two significant modifications during this period. The first change was to surface mount the MC68332 to improve reliability, initially it was installed in a surface mount to through-hole adapter socket. The second change was a switch from a proprietary token bus network to the LON.

Pleased with the results in fuel savings that had been obtained through the winter, the operator of the land-fill requested and received the first production IGEM•1. It is now installed and performing its duties faithfully. Another unit has also been installed and is operational on a pipeline engine in the California desert.

The next phase of the IGEM project will be to increase the capabilities of the IGEM•1 control. The I/O 3 board is presently being productionized, after which the 'Enhanced' and 'Maximum' versions of IGEM•1 will be made available. In addition, more software is planned for the LON interface to the MOE application blocks.

With its flexible hardware capabilities and the LON, a lot of interest has been shown for using IGEM•1 for future engine control and management applications.

CONCLUSION

IGEM•1 is unique in its construction and capability. The ability to program an application using ABLS tools and down-load code directly to the units provides IGEM with the flexibility normally found only in large expensive rack mounted controls. With its advanced LON architecture IGEM•1 can combine many control functions, thus providing a complete integrated gas engine management system.

REFERENCES

Rosauer R.J., 1985, "A Menu Driven Block Oriented Programming Language for Digital Fuel Controls," ASME Paper 85-GT-105

Martin J. with Kavanagh Chapman K., 1989, "Local Area Networks: Architectures and Implementations," Prentice Hall, N.J.

ICE-Vol.18, New Developments in Off-Highway Engines
ASME 1992

TOTAL ENGINE CONTROL REDUCES OPERATIONS COSTS AS WELL AS EMISSIONS

Dwight S. Whiting
Dynalco Controls
Fort Lauderdale, Florida

ABSTRACT

A growing number of users of stationary industrial engines today are concerned primarily about two things: operations costs and emissions restrictions. The use of microprocessors for engine control is becoming prevalent in industry. With the increase in processing power and the corresponding decrease in processing cost, systems capable of total engine control are emerging. Such systems provide increased sophistication and integration. The potential benefits that total engine control offers are reductions in operating costs and harmful emissions because of more efficient and precise engine control and monitoring.

INTRODUCTION

The microprocessor has brought many benefits to a wide range of products. Processing power that once required an entire air conditioned room full of equipment is now available in a single chip that can function comfortably in an engine room.

Today, engine automation is frequently accomplished using one microprocessor, if not several. Simple availability of microprocessor-based products, however, should not be the driving force behind automation. A successful automation program needs a goal. Often, that goal is to do one or both of the following: reduce operating costs and meet regulatory requirements to reduce harmful emissions. Total engine control using microprocessor technology is a means toward both ends.

MICROPROCESSORS IN OUR INDUSTRY

Since its introduction in 1971, the microprocessor's performance has increased geometrically while processing costs have retreated steadily. New capabilities coupled with new ideas have brought entirely new possibilities to both consumer and industrial products. For the engine control market, this means greater precision in data acquisition and control as well as quicker response. It also brings new capabilities such as continuous monitoring, remote data logging, and expanded programmability.

One does not have to look far in the engine room to see a microprocessor-based product. Several dedicated, stand-alone products such as temperature scanners, ignition controls, and air/fuel ratio controls have all found successful applications.

In addition to these, many companies have automated various engine monitoring and control functions themselves using programmable logic controllers (PLCs). The power and flexibility of such microprocessor-based products allow a single unit to take on many functions.

TOTAL ENGINE CONTROL

The problem with having many stand-alone microprocessor-based products is that each has its own power requirements, programming standards, sensor requirements, and enclosure requirements. Each may come from a different vendor or distributor, thus allowing the potential for support problems. Finally, although it may be beneficial to the engine, interaction between stand-alone units is often cumbersome, if not impossible. Interaction can mean grounding problems, calibration nightmares, and protocol translation headaches. Perhaps a PLC or similar unit is the solution.

Most PLCs offer full programmability with a myriad of input/output (I/O) options. The programming package typically includes preprogrammed modules for desirable functions such as PID controllers, timers, comparators, and flow equations. Some functions, however, are difficult, if not impossible to implement. Such functions include real-time, interrupt-driven functions such as crankshaft-referenced ignition, speed governing, knock detection, and pressure versus volume horsepower computation. In such cases, a dedicated total engine control is superior.

The biggest problem with a PLC or similar product is that it typically comes unprogrammed; it must be configured and programmed. Many companies underestimate the expertise and time required to design, program, test, document, and support an engine control system using such equipment.

One solution to the aforementioned problems lies in a relatively new concept: total engine control. A total engine control system as discussed here is an integrated electronic engine control system with a single user interface. Actual control actuation can be, and typically is, pneumatic or hydraulic.

TOTAL ENGINE CONTROL FEATURES

There are both requirements and desirable features of a total engine control if it is to be of maximal utility to the end user. One obvious requirement of a total engine control, by virtue of its name, is that it must be multifunctional. It must also be fully integrated.

An integrated system uses resources most efficiently, thus reducing system, installation, and maintenance costs. It essentially incorporates all processing power, program memory, and I/O termination in a central, unified system. Only one power source and one enclosure are required. Multiple functions share inputs, a single user interface, and a single communications port and protocol. Individual functions cooperate so that, for example, a variable computed by one function can be used to bias the control setpoint of another function. All data are available to all functions. Since the system is integrated, programming is symmetric. In other words, programming one function's variables is no different than programming those of another function in the system. Figure 1 diagrams an integrated total engine control.

To accommodate many different engine types, a total engine control should be modular. This allows the user to incorporate only the functions required for a given application and to not have to pay for unused functions. Modularity also allows expandability so that the user can add functions as desired. For example, new processor or I/O boards can be added to the system to provide new functions, much like new cards can be added to a personal computer. Such a multi-processor approach allows new functions to be added as new technologies emerge — functions that aren't even thought of or cost effective now — thus avoiding instant obsolescence.

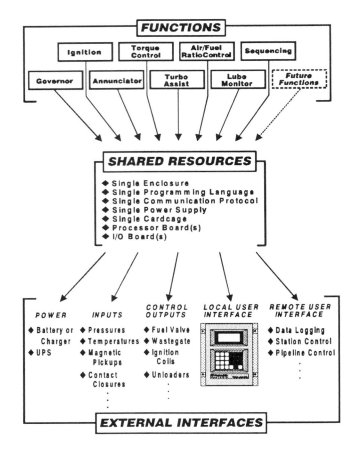

FIGURE 1: TOTAL ENGINE CONTROL

To accommodate many different engine types, a total engine control should also be flexible. It should be easily configured for any engine. It is important to integrate sophisticated control and monitoring capabilities with a simple, intuitive user interface. To reduce the risk of unauthorized programming, the system should also offer multi-level password protection.

Serial communication capability is necessary if remote monitoring and control are desired. Data can be viewed, logged, trended, graphed, or manipulated in other ways. This communication capability is especially necessary if station control or entire pipeline control is desired.

Harnessing some of the processor power for on-line system and engine diagnostics is desirable. Not only can such capability assist in troubleshooting; but it can also provide a degree of engine protection not available with older technology.

Finally, a total engine control system should be designed expressly for the harsh engine room environment where temperature extremes, vibration, moisture, and electrical noise can damage misapplied equipment or cause damage to the engine, and where third party approval is often required. Such requirements are most easily met by an experienced electronic engine controls company. Not only does such a company understand these requirements; but it can also supply such a product at a relatively low cost and in a shorter time frame. All costs of development, documentation, and third party approval are spread over many systems and users; an experienced controls group does not have to be built up for design and support.

BENEFITS OF TOTAL ENGINE CONTROL

Large sums of money can be spent on engine and compressor automation. To be successful, an automation program should have a clear purpose. More often than not, the ultimate goal is to reduce operations costs and maximize revenue. In some areas, the reduction of harmful emissions is an increasingly important goal, since regulatory agencies can force an engine to be shut down. If you can't run the engine, you can't pump the gas!

SAVE MONEY

An integrated total engine control will save money. Initial system purchase price, installation costs, and maintenance costs all are reduced by sharing resources. As discussed earlier, only one power source and one enclosure are required, and input devices can be shared. Speed computed using a single magnetic pickup input can be used for ignition, governing, air/fuel ratio control, torque computation, annunciation, and sequencing. A temperature input can be used to bias ignition timing as well as the air manifold pressure control setpoint. As a result of these efficiencies, labor, conduit, and wiring costs are all reduced.

The sophistication of a microprocessor-based total engine control yields increased operating efficiency as the result of more responsive and precise control. Many engine manufacturers have determined operating curves that when followed will provide the best engine performance. Such performance will typically provide optimal fuel efficiency and lead to reduced maintenance costs over the long run.

An engine's actual operating point may be determined by several inputs. For example, it is not at all uncommon for ignition timing to be based on individual engine speed, air manifold pressure, and air manifold temperature curves. Each curve may be nonlinear, with minimum and maximum limits, and at any time, one curve may take precedence over the others or all the curves may be added together. Temperature may be used only as a bias.

Microprocessor technology excels at such requirements. Figure 2 shows an example of some typical curves. The resultant curve is based on using the lowest of the pressure-based and speed-based timing angles at any time with rate limiting at curve transitions to avoid sudden timing angle discontinuities.

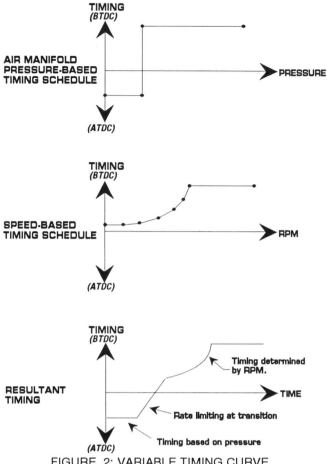

FIGURE 2: VARIABLE TIMING CURVE

With the goal of saving money and maximizing revenue, it is essential to minimize downtime and avoid costly engine failures. As was implied earlier, more responsive and precise control will increase operating efficiency. It follows that a better running engine will have less downtime.

The increased and enhanced diagnostic capability offered by a total engine control will also make it possible to avoid major engine failures before they occur. Such a system can modify control outputs, can alarm, or can shut down the engine to avoid potential problems. An authorized user can program the setpoints and desired action, thus giving the user control over the degree of protection. Conditions are monitored several times a second around the clock.

Serial communications capability also offers cost savings. When an engine alarms, for example, a user can remotely determine what the problem is, rather than take the time to drive to the engine location. Such capability also allows for automated data logging and station or pipeline control. Many companies are doing the latter to improve operating efficiency.

Finally, a total engine control system will have lower support costs. Because it is flexible and modular, one system can be tailored to any engine type. This is especially important when a site is required to stock spare parts and has many different engine types. Training costs are lower. Since one system can run on many engine types, operators need only learn how to operate one type of system. Thanks to symmetric programming, once the operator has learned how to program one function, the others follow easily.

REDUCE HARMFUL EMISSIONS

Many districts, counties, states, and countries are becoming increasingly strict about air quality standards. Such standards are causing many engine operators to take steps to reduce harmful exhaust emissions.

Once again, the sophistication and integration of a total engine control can play a major role in meeting emissions standards. The ability to follow the manufacturer's recommended operating curves under all operating conditions is crucial. Precision and responsiveness are critical. Although it has yet to be the case, future laws may especially prove the benefits of an integrated system. For example, ignition timing, air/fuel ratio, torque, and engine speed may all need to be controlled in a coordinated fashion to reduce a particular exhaust element.

TOTAL ENGINE CONTROL FUNCTIONS

There are many engine control and monitoring functions that may be implemented within a total engine control architecture. The list will grow as sensing, actuation, and processing technologies advance. One product line, the TEC-9000 Total Engine Control, has implemented several functions that are briefly discussed here.

The TEC-9000 offers air/fuel ratio control; flywheel-referenced, capacitive discharge ignition; speed governing; lube monitoring; turbocharger assist control; annunciation (discrete and analog); engine and compressor sequencing; and torque (pocket) control. Each function in the TEC-9000 offers distinct benefits from microprocessor technology, including field-programmability for virtually any application.

CONCLUSION

Total engine control, using today's microprocessor technology, offers distinct benefits to users of stationary industrial engines. Installation, maintenance, and support costs can all be reduced by taking advantage of an integrated total engine control. The precise and responsive control and monitoring made possible by taking advantage of the power and programmability of the microprocessor also lead to increased engine operating efficiency. The end results are then reduced operating costs and reduced emissions.

ICE-Vol.18, New Developments in Off-Highway Engines
ASME 1992

EXPERIMENTS IN TRAP REGENERATION
USING STORAGE OF ENERGY AS HEAT

Alejandro F. Romero
Departamento de Fluidos y Térmica
Instituto de Ingeniería UNAM
Ciudad Universitaria, Mexico

Terry L. Ullman
Department of Emissions Research
Southwest Research Institute
San Antonio, Texas

ABSTRACT

Engine exhaust particulate matter collection using ceramic or fiber filters as inline particulate traps has been successful, as reported in the technical literature. Periodic cleaning or thermal regeneration of these filters by oxidizing the accumulated soot particles using a high temperature flow of air or exhaust gases with sufficient oxygen concentration, however, remains a major problem. This paper presents experimental efforts in obtaining thermal regeneration with a process requiring greatly reduced rates of energy input. The concept of storing energy as heat was used to accumulate energy for delivery at high temperature levels required for regeneration. A large mass of metal, mainly stainless steel, was used as a thermal energy reservoir (TER). Heat stored in the metal mass was then transferred to the particulate trap to obtain controlled thermal regeneration of the trap allowing the trap to continue accumulation of additional exhaust particulate. Experiments with the TER were somewhat successful, but much more work is necessary to produce a viable system for regeneration of particulate traps on buses.

INTRODUCTION

A goal of diesel engine users and manufacturers has been to retain reliable and efficient diesel engine performance, while eliminating smoke, odor, particulate, and oxides of nitrogen. There has long been concern for the carcinogenic potential of diesel exhaust.[1]* Researchers and manufacturers are striving to attain 1994, and later,

particulate emission limits of 0.05 and 0.10 g/hp-hr set by the EPA for urban bus and heavy-duty truck engines, respectively. Simultaneously, oil companies have put substantial effort into producing a low-sulphur diesel fuel ($\leq 0.05\%$ wt.) to reduce particulate emissions.[2,3]

Exhaust aftertreatment devices may be unavoidable in order to meet EPA 1994, and later, federal emission standards for particulate matter emissions. Using ceramic filter media, particulate collection efficiencies on the order of 90 percent, or more, have been reported. The particulate matter collection process has been effective using a variety of filter media in particulate traps. However, regeneration or cleaning of the filter media either by thermal or catalytic means has posed a tremendous challenge. In addition to the technical constraints, traps and their associated regeneration schemes must be developed within cost constraints and must be reliable. Engine manufacturers have expended substantial effort to improve engine combustion and to develop flowthrough catalyst to eliminate the need for a particulate trap.[4]

Initiation of trap regeneration requires high energy levels and temperatures greater than 600°C, unless catalytic coatings or fuel additives are used.[5,6,7,8,9,10] Such exhaust temperatures are seldom encountered in normal bus or heavy duty truck operation. There have been several attempts to use high-power-density devices, like fuel burners or electric heaters. A fuel burner can supply large amounts of power, but it requires complicated controls and very often leads to increase in gaseous emissions.[11] An electric heater can eliminate any direct increase in gaseous emissions, but it requires a high level of electric power, from the 24 V DC battery and generator system, and that system may already be substantially loaded in normal heavy-duty service.

---

*Numbers in parentheses designate references at end of paper.

The thermal energy reservoir (TER) uses the properties of a high-heat-capacity medium to store heat until needed for regeneration. Ideally, the "reservoir" is filled with heat energy over time, such that the rate of input does not tax the existing electrical system on the vehicle. Material contained in the reservoir is heated electrically, using low-power input from the 24 V DC system. The TER is designed for release of heat at a high rate. When the material reaches a pre-determined temperature, the heat energy is transferred to the trap for regeneration. Therefore, the TER represents an electrical regeneration system that ideally eliminates the need of fuel burners and complicated controls. In the trap regeneration literature, it was found that Mazda had filed a patent application on a similar strategy but had not conducted any experiments for proof-of-concept.[12,13] This paper reports some of the experimental results obtained when using energy supplied by the TER heated up to 900°C to regenerate a low pressure drop ceramic filter system in a reverse-flow manner.

EARLY TRAP REGENERATION EXPERIMENTS USING THE TER

For proof-of-principle, a thermal energy reservoir (TER) was constructed of a bundle of stainless steel rods weighing about 50 pounds. Eight specially constructed Watlow Firerods® were inserted perpendicularly through the TER rod bundle to heat the metal mass. The Firerods® were capable of operating up to a temperature of 1800°F (1000°C), and each element was rated for 400 Watts at 24 V DC, for a total of 3.2 kW. Figure 1 shows the TER surrounded with insulation. Piping below the TER was used to route air through the TER and into the ceramic trap element to be regenerated.

Figure 1. Preliminary TER System With Trap Removed

For experimentation, several ceramic filter elements (5.6 x 6.0 inch) were connected to the exhaust of a diesel engine to collect particulate matter. The TER was connected to a charging system and a set of batteries, and the internal temperature of the rod bundle was brought to 900°C. Ambient air was pumped through the TER and then through the trap. For these experiments, the air flow was varied from 20 m^3/h to 60 m^3/h and the direction of flow through the loaded trap elements was reversed.

In most cases, the normal flow direction during trap regeneration is the same as used during collection of soot. Reversing the normal flow direction during regeneration, inverse regeneration, has the advantage of dissipating the heat of soot combustion into the remaining soot particles, instead of into the ceramic substrate. Therefore, inverse regeneration minimizes the thermal shock to the trap ceramic substrate. Inverse regeneration was previously used with partial success using a light-duty vehicle, as reported by Hayashi, K., et al. of Toyota [14].

Although early experiments illustrated the potential viability of the TER system using reverse regeneration of particulate traps, some cracks were detected during the experimentation. Figure 2 shows four regenerated traps in order of regeneration quality from left to right, with the trap on the right displaying the best regeneration, without apparent cracking or other damage (melting). These early results were very effective in identifying problems, such as the large thermal gradients at the inlet of the trap, extreme heat losses, and very long periods of time necessary to heat the TER to sufficient temperature.[15]

Figure 2. The Four Particulate Traps Used in Regeneration Experiments Using Reverse Flow (In Order of Regeneration from Left to Right)

FULL SCALE APPLICATION TO TRANSIT BUS

Using the TER concept, a self regenerating system for particulate matter collection was designed for use mainly as a retrofit in urban passenger buses. The proposed design, shown in Figure 3, consisted of two ceramic filters

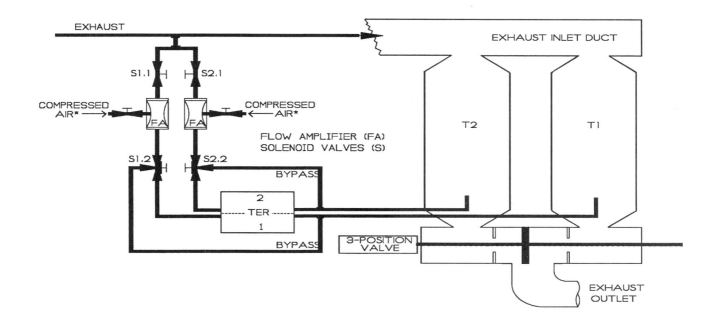

Figure 3. Schematic of Aftertreatment System

for parallel operation (Trap T_1, Trap T_2), a TER, solenoid valves to divert airflow to the TER and to either of the filter elements, flow amplifiers, and a pneumatic three-position valve to switch the regeneration to either filter element.

The system was retrofitted onto a Detroit Diesel 6V92TA engine, configured in a GMC RTS bus power pack. Two ceramic filter elements (11x12) were substituted for the muffler. Figure 4 shows the two trap elements, Traps T_1 and T_2, mounted to the power pack frame. A dual-bed TER was made of stainless steel plates with two independent air passages to provide a means to transfer the stored heat to Trap T_1 or T_2. The TER weighed approximately 40 pounds. Heat energy for storage was supplied to the TER by three rod heaters rated at 600 W at 24 V DC, for a total of power of 1.8 kW.

Ideally, while both trap elements accumulate soot in parallel, the TER accumulates heat energy to a temperature of about 900°C. As the pressure drop across the trap elements increases, the need for regeneration is triggered. The three-position valve switches all exhaust flow from both trap elements to one, blocking exhaust flow to the other. Air (with some exhaust) is directed through flow amplifiers, and through one half of the TER, where it is heated to about 600°C, then the heated air is directed to the exit side of the trap element to be regenerated. The 600°C air stream passes through the trap in a reverse flow direction and proceeds with the engine exhaust out through the other trap element. After regeneration of this trap

Figure 4. Particulate Trap Assembly Mounted in Place of Muffler on a GMC RTS Bus Power Pack

element, the three-position valve is positioned in the opposite position to similarly regenerate the other trap element. After regeneration of both elements, the three-position valve returns to neutral position and exhaust flows through both trap elements, while the TER accumulates heat energy for the next regeneration sequence. Because regeneration was to be performed in reverse flow direction and high temperature air was used, it was envisioned that the regeneration episode would only require two to three minutes, compared to other systems requiring 15 to 20 minutes. Such a relatively short time for regeneration would allow more time between regenerations, enabling slow accumulation of heat energy in the TER using relatively low wattage heater elements.

The components of the trap-regeneration system for the GM RTS power pack were mounted onto a separate fixture, to facilitate experimentation. The traps were loaded to relatively light levels using diesel engine exhaust from a separate engine. The TER was heated to 900°C using a 24 V DC generator and battery system. During early experiments, problems with heat loss were substantial. With additional insulation, 600°C air was provided to the exit face of one of the trap elements. Reconfiguration of the hot-air delivery piping and provisions for a means to diffuse the hot air over the area of the trap face helped to obtain higher temperatures at the exit faces of the trap elements.

Further testing of the full scale system was carried out using a simulated regeneration experiment to identify other problems. With the TER at 900°C, the regeneration process was begun such that air proceeded through the first TER circuit and on to the exit face of Trap T_1. After approximately 2 minutes, hot air at the exit face of trap T_1 reached 600°C and it was maintained for approximately ten minutes, while the TER bulk temperature dropped to about 750°C. Airflow through the first TER circuit was discontinued at 18 minutes when the temperature at the exit face of Trap T_1 decreased to 550°C. The TER bulk temperature dropped to 725°C. After approximately 15 minutes of electric heating, the TER bulk temperature was brought up to 800°C, and airflow was directed through the second TER circuit. The exit face of Trap T_2 also reached 600°C in about two minutes, and peaked at approximately 700°C. As before, the exit face temperature of the trap remained above 600°C for ten minutes, while the TER bulk temperature dropped to about 700°C.

Variations of this experiment were performed, and although temperatures at the trap exit faces were sufficient for regeneration, temperatures within the ceramic substrates did not reach high enough levels to reliably obtain a thorough regeneration. Although no further experiments were performed with the full scale system, potential improvements are being considered because the TER concept promises the use of reduced rates of energy input along with simplified system control to obtain trap regeneration.

CONCLUSIONS

The experimental work performed with the proof-of-principle apparatus and with the full-scale system showed that regeneration temperatures could be obtained using the TER scheme to accumulate and release heat energy, however, hot air mass flow was not sufficient to thoroughly regenerate the trap as necessary for reliable trap regeneration. Space constraints for trap and regeneration hardware pose special problems for containing the high level (high temperature) heat energy needed to obtain regeneration. Routing the heat energy to the necessary location poses additional problems due to difficulties in providing thermal insulation or isolation. Further design work utilizing phase change materials (molten metal) within the TER would help improve the energy storage density. However, improved insulation schemes are needed to substantially reduce heat losses. Ideally the TER system could be designed into a viable system and utilize waste energy from bus braking to replace electrical system load. The search continues!

ACKNOWLEDGEMENTS

This work was performed under the Bus Emissions Technology Cooperative Industry Project, on behalf of its members: Southern California Rapid Transit District (SCRTD), Los Angeles County Transportation Commission (LACTC), Southeastern Pennsylvania Transportation Authority (SEPTA) and Texas Transit Association (TTA). Dr. Romero worked on the TER system at SwRI during a one-year sabbatical leave from the Universidad Nacional Autónoma de Mexico.

REFERENCES

1. Lipkea, W.H., Johnson, J.H., and Vuk, C.T., "The Physical and Chemical Character of Diesel Particulate Emissions-Measurement Techniques and Fundamental Considerations," SAE Paper 78018, Detroit, MI, 1978.

2. Anonymous, "Lubricants and Fuels for Buses of the Future," Lubrizol NewsLine, Vol. 10, No. 1, January, 1992.

3. Springer, K.J., "Low Emission Diesel Fuel for 1991-1994," from Advances in Engine Emissions Control Technology, ICE, Vol. 5, pp. 1-10, ASME, 1989.

4. Khair, M.K., "Progress in Diesel Engine Emissions Control," ASME Paper 92-ICE-14, Houston, TX, January, 1992.

5. Ullman, T.L., Hare, C.T. and Baines, T.M., "Preliminary Particulate Trap Tests on a 2-Stroke Diesel Bus Engine," SAE Paper 840079, Detroit, MI, 1984.

6. Bykowski, B.B., "Formulation and Evaluation of Alternate Diesel Particulate Trap Media," ASME Paper 87-ICE-36, Dallas, TX 1987.

7.   Springer, K.J. "Particulate Trap for Two-Stroke Cycle Detroit Diesel Powered City Bus," from Engine Design, Operation, and Control Using Computer Systems, ASME-ICE, Vol. 9, pp. 9-20, 1989.

8.   Hardenberg, H.O., "Urban Bus Application of a Ceramic Fiber Coil Particulate Trap," SAE Paper 870011.

9.   Barris, M.A., and Rocklitz, G.G., "Development of Automatic Trap Oxidizer Muffler Systems," SAE Paper 890400.

10.   Rao, V.D., et al., "Advanced Techniques for Thermal and Catalytic Diesel Particulate Trap Regeneration," SAE Paper 850014.

11.   Romero, A.F., and Ullman, T.L., "Emissions Testing of a 3208 Caterpillar Diesel Engine Equipped with a Donaldson-Webasto Particulate Matter Filter System," Final Report, SwRI Project 4345, November, 1991.

12.   Hamada, S., Patent 60-128920 "Exhaust Repurifier of Diesel Engine."

13.   Misumi, M., Mazda Motor Corporation, personal communication, 09/02/1991.

14.   Hayashi, K., et al., "Regeneration Capability of Wall-Flow Monolith Diesel Particulate Filter with Electric Heater," SAE Paper 900603.

15.   Ullman, T., Human, D.M. and Carroll, J.M., Progress Report No. 6, 7, and 9 for (Bus Emission Technology Cooperative Industry Project," SwRI Project 08-2481, 1989-1990.

ICE-Vol.18, New Developments in Off-Highway Engines
ASME 1992

CARBONYL COMPOUNDS AND PAH EMISSIONS FROM CNG HEAVY-DUTY ENGINE

M. Gambino, R. Cericola, P. Corbo, and S. Iannaccone
Istituto Motori C.N.R.
Naples, Italy

ABSTRACT

Previous works carried out in Istituto Motori laboratories have shown that natural gas is a suitable fuel for general means of transportation. This is because of its favorable effects on engine performance and pollutant emissions. The natural gas fueled engine provided the same performance as the diesel engine, met R49 emission standards, and showed very low smoke levels.

On the other hand, it is well known that internal combustion engines emit some components which are harmful for human health, such as carbonyl compounds and policyclic aromatic hydrocarbons (PAH). This paper shows the results of carbonyl compounds and PAH emissions analysis for a heavy-duty Otto cycle engine fueled with natural gas. The engine was tested using the R49 cycle that is used to measure the regulated emissions. The test analysis has been compared with an analysis of a diesel engine, tested under the same conditions.

Total PAH emissions from the CNG engine were about three orders of magnitude lower than from the diesel engine. Formaldehyde emission from the CNG engine was about ten times as much as of the diesel engine, while emissions of other carbonyl compounds were comparable.

INTRODUCTION

In metropolitan areas, vehicles are the prevailing source of atmospheric pollution. In many cities the use of private cars is restricted, and consequently, mobility is based on the public transportation system. Therefore, great attention must be paid to reduce emissions from mass transportation vehicles. In this context, the administration of the Tuscany Region of Italy took the initiative to verify the benefits of using CNG instead of Diesel fuel in urban bus engines. For this purpose Istituto Motori carried out a research activity regarding the conversion of a Diesel engine to a natural gas engine. A comparison between performance and pollutant emissions before and after the conversion was conducted. In order to realize maximum benefit from the comparison of the two fuels, the exhaust emission analysis was performed not only on regulated substances (HC, CO, NOx, smoke), but also on some unregulated compounds such as carbonyl compounds and PAH. These compounds, which are considered highly harmful to human health, were identified and quantified by means of a specific sampling and analysis procedure.

CNG ENGINE

The goal of developing a CNG spark ignition engine with low pollutant emissions can be achieved by one of two basic approaches: using three way catalyst technology, which requires the engine to be operated at stoichiometric air-fuel ratio, or using the lean burn technology. The first solution needs a very accurate control of the air-fuel ratio, which is particularly difficult to realize in a methane fueled engine, where the window of catalyst efficiency is narrower than in a gasoline engine [1]. Moreover, the commercial catalytic converters showed low conversion effect on methane [2], which is the main component of total hydrocarbons present in the exhaust gas of a CNG engine. On the other hand, results of several experiments conducted at the Istituto Motori on different types of heavy-duty natural gas engines showed the possibility of meeting emission limits with the lean burn solution [3]. In addition, a lean burn engine generally showed higher thermal efficiency and lower exhaust temperature than a stoichiometrically adjusted engine. For these reasons the lean burn option appeared to be more viable and suitable, and was chosen in the present work. Engine technical characteristics are reported in Table 1.

A disk shaped combustion chamber with two spark plugs has been adopted. The compression ratio was lowered to 9:1, while valve overlap was eliminated in order to prevent the fresh mixture from escaping through the exhaust valve. A liquid cooled turbocharger was also adopted.

TABLE 1. CNG TURBOCHARGED ENGINE DATA.

6 cylinder turbocharged
displacement 9.5 l
bore x stroke 120 x 140 mm
compression ratio 9:1 (16:1)
max power 154 kW @ 1900 rpm (154 kW @ 2050 rpm)
max torque 776 Nm @ 1800 rpm (882 Nm @ 1100 rpm)
max brake efficiency 0.359 (0.390)
electronic ignition with two spark plugs per cylinder
CNG venturi carburetor DELTEC
boost pressure 0.6 bar
valve overlap angle 0 deg (49.8 deg)

NOTE: the values in brackets are referred to the diesel engine.

ENGINE PERFORMANCE TESTS

The two engines (CNG and Diesel) were tested using a Schenck U1-25H hydraulic dynamometer. Inlet air flow was measured with Ricardo Viscous Air Flow Meters, and natural gas flow was calculated with a Tartarini Instromet 782-2X volumetric flow meter.

In spite of the lean mixtures used, the power of the CNG engine was comparable with the diesel version, as shown in Figure 1. On the other hand brake specific energy consumption (BSEC) was slightly higher for the CNG version, due to the lower compression ratio. For exhaust gas analysis the following instrumentation was used:
- Beckman 404 HFID (flame ionization detector) for hydrocarbons;
- Beckman 880 NDIR (non dispersive infrared analyzer) for CO and CO_2;
- Beckman 955 CLA (chemiluminescent analyzer) for NOx;
- Hartridge MK III smoke meter to measure exhaust gas opacity.

Regulated gas emissions were measured according to 88/77 CEE R49 procedure (13 mode cycle), while smoke opacity was evaluated following the R24 procedure.

Pollutants emitted during the R49 cycle were lower than the limits for both engines. In Table 2 the comparison between gaseous emissions of the two engines is reported.

TABLE 2. R49 TEST FOR CNG AND DIESEL ENGINE.

	HC	CO	NOx
		g/kWh	
CNG	2.0	2.3	11.9
Diesel	1.0	1.4	15.3

The high HC emission obtained from the CNG version can be attributed to the combustion chamber shape, which was not specifically optimized for natural gas. However, it is important to mention that the hydrocarbons emitted from the CNG engine were mainly constituted by methane, which is not toxic.

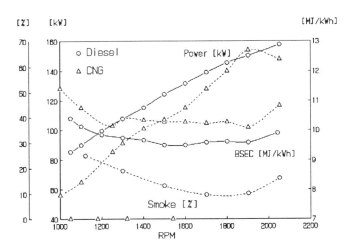

FIGURE 1. POWER, BSEC, AND SMOKE FOR CNG AND DIESEL ENGINE.

On the other hand, the lower NOx emission from the natural gas engine was due to using lean mixtures and also to the different type of combustion. Fig. 1 shows that exhaust opacity is negligible (2% Hartridge) for the CNG engine, while it varies from 10 to 30% for the diesel engine.

Starting in July 1992, stricter emission limits will be imposed in Europe to heavy-duty engines. In order to meet the new limit standards, some improvements of the engine components are necessary, in particular, the combustion chamber shape. In fact, a microturbulence combustion chamber should allow extremely lean mixtures to be used, resulting in very low emissions and good thermal efficiency [4]. In addition, appreciable reduction of pollutant emissions should be achieved by the electronic control of the gas feed system.

UNREGULATED EMISSIONS

Sampling and analysis procedure

The sampling components train is illustrated in Figure 2. Unregulated emissions were collected by two stainless steel probes that were located in the exhaust pipe at 1.5 m from the turbocharger.

The sampling and quantitative analysis of carbonyl compounds were performed by using the 2,4-dinitrophenylhydrazine (DNPH) method [5]. The compounds were collected by bubbling the engine exhaust gas through an acetonitrile solution of DNPH reagent, cooled at 0 ˚C. A filter placed upstream of the impingers stopped the solid particles. The face temperature of this filter was about 80 ˚C. Samples of the solution containing the hydrazone derivatives were directly injected into the Hewlett-Packard high pressure liquid chromatograph, which was equipped with an ultra-violet detector.

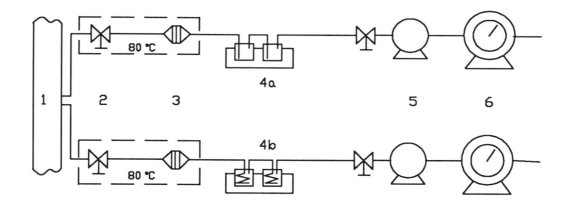

FIGURE 2. SAMPLING TRAIN FOR CARBONYL COMPOUNDS AND PAH COLLECTION:

1	exhaust pipe	4b	condensers
2	two way valve	5	vacuum pump
3	filter	6	dry gas meter
4a	impingers		

The analysis of the carbonyl compounds showed an appreciable quantity of several compounds, some of which are considered significant in the IARC classification [6].

PAH can be present in the exhaust of an engine in both the particulate matter and the gas phase. For this reason, the sampling line was comprised of a Pallflex filter type TX40HI20-WW for soot collection and two condensers assembled in series (0°C and -20°C) to stop vapour phase PAH. The outside temperature of the filter was maintained at about 80 °C. After sampling the line, the condensers were washed with methylene cloride, in order to recover any organic material lost along the line. The condensed water was separated by phase separator filters. Soxhlet extraction with methylene chloride was performed on the filter. The organic material contained in the extract was added to that which was trapped in the condensers, then all the solvent was roto-evaporated. The total organic material obtained by the described procedure was analyzed by a Hewlett-Packard GCMS (gas chromatography-mass spectrometry) using the SIM technique (selected ion monitoring). According to EPA method 610 and IARC classification, only 16 PAH were identified and quantified. Detecting this group of compounds is a significant indicator of the toxicity grade of engine emissions.

The unregulated emissions were evaluated using the same 13 mode cycle used for regulated emissions.

Test results and data analysis

A summary of unregulated emission results is given in Table 3 and 4. PAH emission data (Table 3) show a significant difference between CNG and Diesel engines.

TABLE 3. POLICYCLIC AROMATIC HYDROCARBONS (PAH) EMISSIONS ON 13 MODE CYCLE (R49)

	CNG		Diesel
		$\mu g/kWh$	
Naphthalene	$1.76 \cdot 10^{-1}$		$9.95 \cdot 10^{+1}$
Acenaphthylene	$1.63 \cdot 10^{-1}$		$1.62 \cdot 10^{+2}$
Acenaphthene	$5.61 \cdot 10^{-2}$		$8.28 \cdot 10^{+1}$
Fluorene	$1.80 \cdot 10^{-1}$		$2.52 \cdot 10^{+2}$
Phenanthrene	$8.65 \cdot 10^{-1}$		$1.03 \cdot 10^{+3}$
Anthracene	$1.74 \cdot 10^{-1}$		$7.63 \cdot 10^{+1}$
Fluoranthene	$2.70 \cdot 10^{-1}$		$6.70 \cdot 10^{+1}$
Pyrene	$2.90 \cdot 10^{-1}$		$1.19 \cdot 10^{+2}$
Chrysene	$9.44 \cdot 10^{-2}$		$6.75 \cdot 10^{+0}$
Benz(a)anthracene (2A)	$2.60 \cdot 10^{-1}$		$2.13 \cdot 10^{+1}$
Benzo(b)fluoranthene +			
Benzo(k)fluoranthene (2B)	$6.09 \cdot 10^{-2}$		$4.53 \cdot 10^{+0}$
Benzo(a)pyrene (2A)	$1.08 \cdot 10^{-1}$		$3.90 \cdot 10^{-1}$
Indeno(1,2,3-cd)pyrene (2B)	$1.09 \cdot 10^{-2}$		$5.61 \cdot 10^{-1}$
Dibenz(a,h)anthracene (2B)	$7.43 \cdot 10^{-3}$		$2.90 \cdot 10^{-1}$
Benzo(ghi)pyrene	$1.18 \cdot 10^{-1}$		$6.23 \cdot 10^{-1}$
Total Policyclic Aromatic Hydrocarbons	$3.38 \cdot 10^{+0}$		$1.92 \cdot 10^{+3}$

PAH emission data of the Diesel engine are in a good agreement with the work conducted by Hemmerlein et al. on the same 13 mode cycle [7].

Table 3 shows that production of the PAH group from the Diesel engine was three orders of magnitude higher than from CNG engine. In particular, differences from 1 to 3 orders of magnitude can be noticed for those PAH to which correspond a higher toxicological risk in IARC classification: benz(a)anthracene, benzo(a)pyrene and dibenz(a,h)anthracene which are present in Group 2A of such a classification, benzo(b)fluoranthene and indeno(1,2,3-cd)pyrene, included in Group 2B.

Concerning total carbonyl compounds, data of Table 4 show that emissions from the CNG engine are almost an order of magnitude higher than from the Diesel engine.

TABLE 4. CARBONYL COMPOUNDS EMISSIONS ON 13 MODE CYCLE (R49)

	CNG	Diesel
	mg/kWh	
Formaldehyde (2A)	32.48	3.83
Acetaldehyde (2B)	1.29	1.82
Acrolein (3)	0.32	0.30
Acetone	1.76	0.58
Propionaldehyde	0.18	0.12
Crotonaldehyde	0.12	0.42
Methacrolein	0.11	0.07
Butyraldehyde +		
2-methyl ethyl ketone	0.20	0.15
Benzaldehyde	0.30	0.57
Total carbonyl compounds	36.76	7.86

This difference is mainly due to emission of formaldehyde (Group 2A in IARC classification), which can be considered a typical emission of a methane fueled engine operating in lean burn conditions [4]. The other compounds included in the IARC classification, showed that a lower emission of 30% for acetaldehyde (Group 2B) was obtained from the CNG engine, while emission values of acrolein (Group 2B) measured for the two engines were comparable.

In order to evaluate the difference observed for the unregulated emission values of the two engines and its impact on health, the relative toxicity of the compounds included in Group 2A of IARC classification was estimated. A relative toxicity factor was calculated using the threshold limit values (TLV) established for concentration of harmful substances in workplace. In particular, the TLV used for formaldehyde is that fixed by the American Conference of Governmental Industrial Hygienist (ACGIH) [9]. It is defined as the time weighed average concentration (TWA) of harmful substance, to which most workers can be exposed without negative effects on health. However, as ACGIH does not establish any TLV for PAH, the value of maximum allowable concentration (MAC) fixed in Russian regulation and suggested in France for benzo(a)pyrene [10-12] was used for the three PAH included in

Group 2A of IARC classification. Assuming TLV and MAC values as a measure of the relative toxicity of the considered compounds, the relative toxicity factor for PAH is four orders of magnitude higher than for formaldehyde, as shown in Table 5.

TABLE 5. RELATIVE TOXICITY OF UNREGULATED EMISSIONS

	HCHO	PAH*
TLV-TWA	1.5 mg/m³	
MAC		0.15 µg/m³
Relative toxicity factor	1	10^4
13 mode emission		
CNG	32.5 mg/kWh	0.38 µg/kWh
DIESEL	3.8 mg/kWh	21.98 µg/kWh
Toxicity index		
CNG	32.5	4
DIESEL	3.8	220
Total index (HCHO+PAH*)		
CNG	36	
DIESEL	224	

* Benz(a)anthracene+Benzo(a)pyrene+Dibenz(a,h)anthracene

As a consequence, taking into account the emission values as calculated on the 13 mode cycle, it is possible to estimate a toxicity index from the emission value for the relative toxicity factor. Table 5 shows that the total toxicity index, calculated for unregulated compounds which are included in Group 2A of IARC classification, is about one order of magnitude lower for the CNG engine than for the Diesel engine.

CONCLUSION

The CNG engine has proved to be a promising solution for reducing pollutant emissions from urban transit vehicles. In fact, the CNG engine tested in this work (a bus Diesel engine converted to work with CNG) showed very low smoke. CO, HC and NOx emissions were lower than R49 regulation limits. Total hydrocarbons was mainly composed of harmless methane.

Furthermore, the toxicity index, evaluated on the base of PAH and carbonyl compound emissions, is about one order of magnituide lower for the CNG engine than for the Diesel version.

For the lean burn CNG engine to meet the new European emission standards, new designs are needed for both the combustion chamber and the gas feed system.

REFERENCES

[1]   J.Klimstra, "Catalytic Converters for Natural Gas Fueled Engines - A Measurement and control Problem" SAE Paper 872165 (1987).

[2]   M.Gambino, S.Iannaccone, A.Unich, "Performances and emissions of heavy duty spark ignition engines fueled with methane", International Conference on Natural Gas and Liquefied Petroleum Gas as Fuel for Internal Combustion Engines, Kiev, September 1991.

[3]   M.Gambino, S.Iannaccone, A.Unich, "Heavy duty spark ignition engines fueled with methane", Transaction of the ASME, Journal of Engineering for Gas Turbines and Power, Vol. 113, No.3, 359-364 (1991).

[4]   M.G.Kingston Jones, D.M.Heaton, "Nebula Combustion System for Lean-Burn Spark Ignited Gas Engines", SAE Paper 890211 (1989)

[5]   F.Lipari, S.J.Swarin, "Determination of Formaldehyde and Other Aldehydes Automobile Exhaust with an Improved 2,4-Dinitrophenylhydrazine Method", J.Chromatogr., Vol. 247, 297-306 (1982).

[6]   IARC Monographs on the Evaluation of Carcinogenic Risks to Humans, Vol. 46 (1988).

[7]   N.Hemmerlein, V.Korte, H.Richter, "Investigations Concerning Use of Rapeseed Oil as Alternative Fuel for Diesel Engine", Technical Report, Porsche R&D Centre.

[8]   D.Grechi, A.Poggi, V.Bellini, "Da Gasolio a Metano: Effetti sulla Concentrazione Atmosferica di Inquinanti Tossici per l'Uomo e sulla Rumorosità (From Gasoil to Methane: Effect of Toxic Pollutants on Human Health and Noise)", Conference on Using of Natural Gas for Public Transport, Florence, Italy, May 1992.

[9]   Giornale degli Igienisti Industriali (Journal of Industrial Hygienists), Threshold Limit Values ACGIH 1986-1987, Vol. 12, 1 (1987)

[10] L.M.Shabad "On the So-Called MAC (Maximal Allowable Concentrations) for Carcinogenic Hydrocarbons", Neoplasma, Vol. 22, 459-468 (1975)

[11] "Occupational Exposure Limits for Airborne Toxic Substances", Occupational Safety and Health Series, No.37, International Labour Office, Geneva (1980)

[12] "Valeurs Limites d'Exposition Professionelle aux Substances Dangereuses en France (Limit Values of Professional Exposure to Dangerous Substances in France", INRS, Cah. Notes Docum. No.133, 691-706, (1980)

71

A NUMERICAL STUDY OF DIESEL ENGINE EMISSIONS
FROM COMBUSTION OF VARIOUS FUELS

Ismail Celik, Yi-Zun Wang, Bruce G. Crawford, Jr.
and Donald W. Lyons
Department of Mechanical and Aerospace Engineering
West Virginia University
Morgantown, West Virginia

ABSTRACT

The emissions from a typical 4-stroke diesel engine is studied using numerical simulations. Two different computer codes are utilized for this purpose; one is a multidimensional code which account for the mixing process, the details of spray dynamics, and partial chemical kinetics. The other is a zero-dimensional transient code which uses a well stirred reactor assumption with equilibrium chemistry. Various liquid fuels are considered. The results from two different codes are compared with each other as well as with experimental results. The results from the axisymmetric code compares favorably well with the three dimensional calculations and with experiments. Among four types of liquid fuels, ethanol seems to to yield highest levels of NO_2 and diesel the least. Although the global reactor parameters can be predicted reasonably well with the zero dimensional model, the use of equilibrium chemistry and empirical burning formulas cause the emission predictions, in particular NO_x, to be considerably different from experimental as well as from the multidimensional model results.

INTRODUCTION

The pollutants emitted from burning of fossil fuels in diesel engines are believed to be a major contributor to the problem of air pollution which ,recently, has aroused a great deal of social, economical, and political concern. The oxides of nitrogen such as NO and NO_2(or NO_x in general), carbon monoxide, CO, and unburned hydrocarbons, HC, are the major constituents of pollutants emitted from diesel engines. Various approaches have been developed for reducing emission levels (see e.g. Flanagan and Gretsinger, 1992)of these pollutants, however, the results are not sufficient in many cases to satisfy the limits of air quality standards.

One viable alternative for reducing levels of pollutants is to utilize different fuels and fuel blends. This, however, requires a good understanding of the combustion processes in diesel engines which is, in principle, unsteady, turbulent and multidimensional. It is controlled by many parameters, in particular the mixing mechanism of fuel and air in the combustion chamber. The fuel characteristics such as composition, viscosity, and droplet size play an important role in determining the nature of combustion process, and hence the formation of pollutants. Experimental research is limited in understanding and predicting performance and emission characteristics because of the aforementioned complexities. Measurements during a real transient operation mode are very difficult, and in some cases impossible. It is a formidable task to measure all of the large number of constituents in the exhaust gas. For example the NO_x formation is believed to be controlled by high speed reactions in fuel rich zones within the flame front(Fenimore, 1971) in addition to the Zeldovich mechanism. Thus quenching and then using gas analyzers may not be a reliable technique for detecting such species. On the other hand computer simulations, when properly validated and verified, do not have these disadvantages or limitations such as high temperatures and transient operation conditions. They can be used to study a large number of cases by varying parameters such as fuel type and fuel additives to predict and control pollutant formation. The validation and verification procedure is a key element in building confidence on predictions. It must also be pointed out that mathematical models still use empirical information at some point or another. Thus, the computer simulations will be meaningful only if they are run in parallel with experimental research to supplement each other.

The computational models relevant to internal combustion engines are reviewed in detail by Gosman et al. (1980),Amsden et al. (1985), O'Rourke and Amsden (1987), Ramos(1989), and by Kamimoto and Kobayashi(1991). The models vary from zero dimensional-transient models to three dimensional transient models with many variants of one dimensional and/or zonal models in between. The way that the chemical reactions are handled also varies significantly from one model to another with an increasing degree of complications starting from equilibrium chemistry to full kinetic treatment of a large number of reactions pertaining to the specific fuel used. The effect of turbulence on chemical reaction rates is still an unresolved issue (see e.g. Gosman et al., 1980; Ramos, 1989). Our purpose here is not to go into the details of combustion modeling problems in general, but rather, here we would like to investigate problems that the users face in applications of certain models. Furthermore, we compare results from very simple models with those from fairly complicated models and make an assessment of the relative use of these models from an engineering application perspective.

Among the numerous number of commercial and public domain computer codes suitable for diesel engine simulations we found that KIVA-II (Amsden, et al., 1989 and 1985) is the most commonly used multi-dimensional code. The ZMOTTO of Zeleznik and McBride(1985), which employs a zero dimensional transient model, was selected as a reference code at the basic level.

In this paper we apply these two codes to a typical four stroke spark ignition engine running at a fixed load condition with various fuels, and compare the results with each other as well as with experiments. Only axisymmetric calculations were performed with KIVA. The full tree dimensional calculations with a moderately fine mesh of 10,000 nodes (O'Rourke and Amsden, 1987) take about three hours of CRAY-XMP super computer time for only two strokes of the cylinder motion. We find this still too long which is not within the reasonable time requirements for engineering applications. We assess our results in comparison with the full three dimensional transient results of O'Rourke and Amsden and draw conclusions with respect to the feasibility of axisymmetric and even zero-dimensional models.

GENERAL FEATURES OF THE COMPUTER CODES

KIVA-II is a multidimensional arbitrary-mesh, finite difference hydrodynamics program suitable for internal combustion engine applications. It solves the three-dimensional, unsteady equations of motion of a chemically reactive mixture of ideal gases. It also solves for dynamics of a liquid spray and accounts for the coupling between the gas phase and the liquid phase due to mass, momentum and energy exchanges. Droplet evaporations is calculated empirically using the locally calculated flow and temperature field. A stochastic technique is used to calculate droplet statistics including the effects of droplet breakup, collisions and coalescences. Chemical reactions are divided into finite-rate and equilibrium reactions, and continuity equations are solved for the species which are controlled by finite rate reactions. A standard k-ε turbulence model is used for calculating gas phase turbulence stresses. The effects of turbulence/chemistry intraction are neglected. Chemical conversion rates are given by Arrhenius expressions

where turbulent mean flow quantities are used in place of the laminar values. This seems to be one of the deficiencies of the formulation used in KIVA. The wall heat loss is taken into account via a semi-empirical model using the law-of-the-wall velocity profiles. Further details about the code can be found in Amsden et al.(1985, 1989); O'Rourke and Amsden(1987).

For the present calculations, the cylinder bowl arrangement is approximated by an axisymmetric geometry. To match the actual volume of the bowl the piston head was shortened by 0.15 cm. A swirl velocity is imposed initially and at the circumference using the empirical formulas given by O'Rourke and Amsden (1987). These formulas describe the azimuthal component of the velocity (i.e. the swirl velocity) as a function of swirl number, engine angular velocity and the radial distance from the cylinder axes. Swirl number 6.5 is used. A numerical mesh of 21x23 (radial, axial directions) was used when the piston was at the BDC. This grid was found to be adequate for numerical accuracy in previous calculations of O'Rourke and Amsden (1987).

ZMOTTO calculates cycle performance parameters as well as working fluid compositions and properties for a number of consecutive cycles and for a variety of input parameters. It uses a zero-dimensional transient model in which the conservation of mass, momentum and energy equations are solved for, along with the species concentrations in a perfectly mixed reaction environment. Hence, the spatial variations of flow and species properties inside the engine are neglected all together. The resulting equations are non-linearly coupled ordinary differential equations or algebraic equations. The code can handle a variety of situations with various levels of sophistication in the chemistry sub-model and in describing the flow through the valves. The present study uses Modeling Level 2 which includes heat transfer, finite-rate burning, and equilibrium chemistry with simplified valve dynamics. The calculations start with residual gases from the previous ideal cycles with equilibrium chemistry, constant volume combustion, and no heat transfer.The fuel burning formulas are derived from a general class of Fourier burning formulas, a special case of which is the cosine burning function used by Blumberg and Kummer(1971). It this formula which is used for the present study. The reader is referred to Zelezink(1976), and Zeleznik and McBride(1985) for more details of this computer program.

ENGINE GEOMETRY AND OPERATION CONDITIONS

The engine geometry is based on the experimental (Lewis et al. 1981, 1983 and 1986) and the computational work by O'Rourke and Amsden (1987). The experiments were performed at General Motors Research Laboratories. Table 1 and 4 provide the necessary information for the numerical simulations. The fuel used in the experiments was Amoco91 type which was approximated as n-octane in the calculations. The engine speed was 2,000 rpm for all cases.

Table 1. Engine geometry and spray injection parameters

Engine geometry	
Bore	9.843 cm
Stroke	10.46 cm
Connecting Rod	17.186 cm
Bowl Volume	54.8 cm^3
Bowl Diameter	5.05 cm
Squish Height	0.13 cm
Ring Land Volume	1.6 cm^3
Head	flat
Compression Ratio	13:1
Spary injection parameters	
Injection velocity	8336 cm/s
Sauter mean radius r_{32}	3.9 μm
Cone angle	13°

SPECIES AND FUEL SPECIFICATIONS

Computer code KIVA-II includes 12 chemical species; 1 fuel and 11 other reactants and products. The input file contains kinetic and equilibrium reaction rate data for 10 reactions. The kinetic reactions are:

$$2C_8H_{18} + 25\ O_2 = 16\ CO_2 + 18\ H_2O$$
$$O + N_2 = NO + N$$
$$O_2 + N = NO + O$$
$$OH + N = NO + H$$

and the equilibrium reactions are:

$$H_2 = 2H \qquad O_2 + H_2 = 2OH$$
$$O_2 = 2O \qquad O_2 + 2H_2O = 4OH$$
$$N_2 = 2N \qquad O_2 + 2CO = 2CO_2$$

The fuel combustion is prescribed by an Arrhenius type reaction. The reactions rate parameters used for various fuels (Westbrook and Dryer, 1984; Vargaftik, 1975) are listed in Table 2.

Table 2. Reaction rate parameters used in the calculations

$$K = [Fuel]^a [Oxidizer]^b A T^n\ e^{(-E/RT)}$$

Fuel	A	a	b	E kJ/kg
Octane	8x10^{10}	0.25	1.5	30
Diesel#2	8x10^{10}	0.25	1.5	30
Benzene	2x10^{11}	-0.1	1.85	30
Ethanol	1.5x10^{12}	0.15	1.6	30

RESULTS AND DISCUSSION

We first compare the predictions from the axisymmetric model (KIVA-II) with those from three-dimensional(3D) calculations of O'Rourke and Amsden(1987) using the same code, and the results from the present zero-dimensional(OD) model. We also compare the calculations against measured data for the same case (Case 3, Table 4). A summary of the results for this case is given in Table 3. The pressure distribution for this case is shown in Figure 1. The maximum pressure is predicted by the axisymmetric model at about 10 deg (ATDC) as 3140 kPa(Fig.1), which is in very good agreement with experiments (Table 3). The peak pressure is predicted as 4100 kPa by the 3D model, and as 4218 kPa by the zero-dimensional model. At this stage we do not know the causes for the differences between the different models. However, we found that minute differences in the reactor volume, and the injected fuel mass cause significant changes in the pressure distribution. Also the fuel burning formulas used in the OD model can change the pressure distribution significantly (see Figure 2). The actual physical model used in the experiments has a combustion chamber which is connected to several small volumes usually called crevices. Gas and fuel mixture flow into these volumes during the engine operating cycle. According to Heywood(1986) the piston and ring crevices are the dominant contributor to gas leakage into the crankcase; this is called blowby. One would infer from this that the effective volume of the experimental chamber should be always larger than the apparent volume used in the calculations, therefore leading to a lower pressure in experiments compared to the computations.

The emission predictions from different models are compared with experiments in Table 3. For all models the combustion efficiency, η=(fuel mass burned)/(total fuel mass), is about 100% as opposed to 89% in the experiments. Therefore the predicted CO_2 levels, [Final (CO_2)-Initial(CO_2) = 38 mg], are the same for all the models which is higher than the measured value, as it should be. The axisymmetric model predicts about the same level of NO as the 3D-Model which is two times less than the experimental value. The zero-dimensional model which uses equilibrium reactions gives significantly higher NO concentration compared to other models.

Table 3. Comparison of different models:

	P_{max} KPa	NO mg	CO * mg	CO_2 mg	H_2O mg	Combustion Efficiency % (mass burned)
Experiment	3300	0.0994	-	24.8	24.6	89.2
3-D Model (KIVA-II)	4100	0.05 ** (0.01)	0.48 (0.068)	48.0 (10.04)	- (6.272)	-
Axisym. mod. (KIVA-II)	3140	0.052 (0.0)	0.77 (0.0)	47.8 (10.04)	23.8 (6.272)	98.78
Zero-D (Zmotto)	4218	0.379 (0.0364)	$\approx$ 0. (0.)	42.5 (4.059)	35.61 (3.4)	100.0

* Maximum values at variable crank angle, the other values refer to the end of expansion stroke.
** Values in parentheses are the initial conditions.

The comparison of the peak CO values that forms during the compression and expansion strokes show that the one from the axisymmetric model is much higher than that of 3D-model. The OD-model predicts practically no CO. It should be noted that due to the incomplete combustion and the presence of other species such as hydrocarbons (HC) in the experiments, there is a considerable amount (.66mg) of unreacted CO left. This value, which most probably was measured at the exhaust as an average value, is higher than even the peak values predicted by the models. This can be explained by the fact that in KIVA empirical laminar flow kinetic rates are used for the fuel oxidation reaction. Similar empirical burning formulas are used in ZMOTTO. It seems that these reaction mechanisms are not adequate to describe the actual combustion process which is most probably controlled by the turbulent mixing of fuel and air inside the cylinder chamber.

Calculations with the Zero-Dimensional Model:

As it was mentioned above the mixing process inside the cylinder influences the combustion significantly. In the OD model the only way that the influence of mixing can be accounted for is through the empirically prescribed mass-burning formulas. We compared the calculated pressure variations which resulted from using several different mass-burning formulas in Figure 2. These formulas can be found in Zeleznik and McBride(1985) which are given in a general Fourier series format. It is seen that indeed the pressure distribution inside the cylinder changes significantly from one burning formula to the other. The way that the burning takes place as a function of the crank angle not only changes the maximum attainable pressure but also the angle at which it occurs. This ,of course, is a direct consequence of the variation of combustion heat release as function of the crank angle. Without combustion heat release the maximum pressure should occur at the TDC (i.e. $\theta=0$.) Comparison with the measured pressure distribution in Figure 1. shows that Formula II with Fourier Coefficients $a_0=3/8$, $a_1=-1/2$, $a_2=1/8$, and $a_3=0$. gives the best agreement between calculations and experiments. On this ground this formula was used for all other cases.

Another important parameter in this model is the heat transfer coefficient which is also prescribed empirically. Of the many available options we found that the following form worked reasonably well for the presently studied engine:

$$h = 0.32 \ Pr^{0.4} \ Re^{0.8} \qquad (1)$$

where Pr is the Prandtl number, and Re is the Reynolds number based on the a velocity calculated as a function of the piston speed, angular velocity of the engine, and the fuel-mass burning rate.

Three cases with different air/fuel ratios were simulated with the OD model keeping the model parameters fixed. The fuel is n-octane and the engine is the UPS-292 stratified charge engine which was described above. The calculations were started from an ideal cycle and continued for 5 cycles. The recirculated mass percentage from the previous cycle was 5%, and the relative humidity of intake air was 100% for all cases. The key parameters as predicted by the model are compared with measurements(Lewis et al. 1981, 1986) in Table 4 and 5. Comparison of these tables show that the trends in combustion efficiency, heat loss ,exhaust temperature, and maximum pressure are calculated in accordance with experiments. As the air/fuel ratio increases from 22% to 61% there is a significant decrease in all these quantities. The quantitative agreement between the predicted and measured values of these parameters is also fairly good. The calculated peak pressures are approximately 30 to 40 % percent higher than the measured ones. The KIVA-3D predictions of O'Rourke and Amsden(1987) where the calculated pressures were higher, but the best agreement was in Case 3. The most important parameter influencing the pressure was found by O'Rourke and Amsden to be the pre-exponential factor used in the reaction rate expression for the fuel and air. This confirms our sensitivity analysis with mass burning formulas as discussed above. The heat loss from the engine has also some effect on the pressure, but calculated heat loss from the OD model is in reasonably good agreement with measurements the calculated values being on the high side for case 1 and 2; this trend was the opposite for KIVA-3D

calculations. Hence the heat loss can not be the cause of the disagreement for predicting higher pressures. Therefore the history of fuel burning which is a strong function of turbulent mixing and spray dynamics need to be prescribed more accurately for better predictions.

Table 4. Experimental conditions and results (Lewis & et al.)

Cases	1	2	3
mass of fuel injected (mg)	31.3	20.8	12.3
mass trapped (mg)	807.4	847.6	849.5
injection & ignition timing(BTDC)	26	19.3	17.5
overall air-fuel ratio	22.4	35.2	61
liner temperature (K)	390	380	369
head temperature (K)	406	394	379
piston temperature (K)	430	418	400
combustion efficiency	98.2	96.1	89.2
heat loss (J)	213.2	161.2	132.6
exhaust temperature (K)	933	703	538
exhaust valve temperature (K)	1200	875	600
pressure Max. (Kpa)	5450	3800	3300
crankangle at pressure max.(ATDC)	10	10	10

Table 5. Computational conditions & results for OD model

Cases	1	2	3
mass of fuel (mg)	31.39	20.95	12.37
mass trapped (mg)	804.21	836.09	847.84
Injection & ignition timing(BTDC)	26	19.3	17.5
overall air-fuel ratio	22.4	35.2	61
liner temperature (K)	390	380	369
mass efficiency	0.975	0.968	0.958
heat loss (J)	358.93	218.56	125.18
exhaust temperature (K)	599.98	524.31	443.51
pressure max. (KPa)	7767	5489	4218
Crankangle at pressure max.(ATDC)	6.52	10.95	8.72

Table 6. Key simulation parameters and calculated species masses

Fuel	m_f mg	ϕ init	P_{max} Kpa	T_{max} K	CO_2 mg	CO* mg	NO mg	H_2O mg	unburned fuel mg	η combustion efficiency
n-Octane	12.3	0.25	3140	2425	47.8	0.77	0.052	23.8	0.1499	98.78%
Diesel#2	12.3	0.24	2783	2174	33.7	0.048	0.005	14.3	4.072	66.89%
Benzene	12.3	0.15	3710	2539	51.4	0.947	0.129	15.0	0.158	98.72%
Ethanol	12.3	0.22	3684	1686	31.2	0.635	0.523	12.2	0.00008	100%

* Maximum values at variable crank angle.
** Air/fuel ratio = 61 for all cases.

Results from Alternate Fuels:

Table 6 summarizes the key simulation parameters used and the calculated emission levels from four different liquid fuels, namely, n-Octane (C_8H_{18}), diesel#2 ($C_{10.8}H_{18.7}$), benzene(C_6H_6), and ethanol(C_2H_5 OH). The total fuel mass injected over a cycle, m_f, and the air to fuel ratio, A/F=61, were kept fixed for all cases,for brevity. All of the mixtures were on fuel lean side, the one with ethanol being the leanest. The fuel injection nozzle parameters and spray model were not changed from fuel to fuel. Since neither the combustion efficiency nor the overall thermal efficiency are not known a priori, it is difficult to predict which fuel would give the maximum torque output in advance. Hence it was assumed that it was possible to run this engine at a speed of 2000 rpm with each of these fuels at a minimum fuel injection of 12.3 mg per cycle.

The pressure histograms depicted in Figure 3 show that the two relatively lighter fuels ethanol and benzene give the maximum pressure. It should be noted that the pressure values shown in this figure correspond to a moving point roughly at the center of the combustion chamber. The pressures for the relatively heavier fuels n-octane , and diesel are considerably lower. An examination of the temperature history(not shown here) indicated that the maximum local temperatures were about the same ($\cong$2500K) for octane and benzene. The peak temperature was somewhat lower for diesel ($\cong$2200K), and for ethanols it was the lowest ($\cong$1700K). There was a considerable ignition delay(=27.5°) in case of diesel. Benzene and ethanol seemed to ignite almost instantaneously. When the ignition delay is long the maximum compression does not correspond to the time of maximum combustion heat release, hence leading to lower pressures inside the cylinder. As for ethanol, it has a considerably low heating value(=27kJ/g) compared to the other fuels(40-44 kJ/g). This would lead to lower temperatures. The pressure is still high because the combustion heat release takes place immediately after the maximum compression.

Comparison of CO_2 production from different fuels(Table 6, Figure 4) shows that the same engine emits the highest CO_2 when it is running with benzene. However, this should not be misleading, for that in case of benzene the combustion is almost complete (η=98.78%, Table 6), whereas for diesel η=67%. The CO_2 emission from burning of ethanol seems to be much lower compared to the other fuels which is consistent with its carbon content. Looking at the unburnt fuel amounts one can infer that soot formation and various HC emissions should be much higher in case of diesel. The CO_2 levels at the end of one cycle is, of course , proportional to the carbon content of the fuel and the amount of fuel burned (i.e. the combustion efficiency). Using these variables a carbon balance calculation was performed which showed that carbon was conserved to a very good accuracy in the calculations.

Figure 4 shows that the maximum CO during the cycle is formed when burning benzene. One can speculate then that with benzene CO emissions should be higher compared to other fuels. It is interesting to see that the peak CO formation with diesel is very low. As for NO, Figure 4 shows that emissions are highest for ethanol, and the lowest for diesel; those for benzene and octane being in the middle. The reaction temperature being relatively low for the case of ethanol, it is surprising to see that this case results in the highest NO emissions.

Emissions calculated with the OD model

The emissions from the same engine using n-octane, benzene, and ethanol were also calculated using the OD model. The results from n-octane were discussed above. The CO_2 levels match with the KIVA results if the mass-combustion efficiency is 100%. Practically no CO is predicted by the OD model using equilibrium chemistry. The NO level is very high for Benzene(0.27mg), and very low(0.08mg)for ethanol. This must be because in reality the NO formation is controlled by kinetics rather than equilibrium reactions as was the case in the OD model.

CONCLUSIONS

The combustion process and the resulting emissions from a typical 4-stroke engine operating at a fixed load have been studied using numerical simulations. Two different readily available computer codes were used, namely, KIVA-II and ZMOTTO. Predicted emissions from combustion of four different fuels are compared with each other. The results of this study lead to the following conclusions:

The axisymmetric model predictions compare favorably well with the full three dimensional calculations previously reported in the literature for the same engine. Both set of calculations are generally in good agreement with experimental results. Hence, for economical reasons, the axisymmetric model can be used instead of the full 3D calculations. The predicted CO levels are significantly lower compared to measured ones. This may be due to faster burning rates seen in the predictions where the reactions are controlled by chemical kinetics rather than turbulent mixing, i.e, the influence of turbulence on kinetic reaction rates are not taken into account.

The zero-dimensional model seems to predict the global combustion parameters such as reactor pressure , temperature, and wall heat loss, fairly well, but when used with equilibrium chemistry it fails to predict the levels of CO and NO emissions accurately. If this model is calibrated against experimental data, and used with appropriate chemical kinetics, it may prove to be a useful engineering tool for quick and inexpensive parametric analysis.

Among the four different fuels considered, namely, n-octane, diesel#2, benzene, and ethanol, our predictions indicate that combustion of ethanol results in the highest levels of NO, and that of diesel in the lowest. However, this may be a direct consequence of the reaction mechanisms and the kinetic rates incorporated in the model. This should be investigated further in comparison to experimental data.

ACKNOWLEDGEMENT

This work was sponsored by the U.S. Department of Energy (DOE), Office of Transportation Technologies , under contract number DE-FG02-90CH10451. The computations was performed using the WVNET (State of West Virginia Computer Network). The authors wish to thank the DOE program managers: A. Chesnes, J. Allsup, J. Russell, and J. Garbak. They also thank N. Clark and R. Bata for their valuable discussions and input concerning the engineering aspects of these calculations.

REFERENCES

Amsden, A.A., O'Rourke, P.J., and Butler, T.D. (1989) "KIVA-II: A computer Program for Chemically Reactive Flows with Sprays," Los Alamos National Laboratory Report LA-11560-MS, Los Alamos, New Mexico 87545. Also US DOE Report DE89 012805.

Amsden, A.A., T.D. Butler, P.J. O'Rourke and J.D. Ramshaw, (1985) "KIVA - A Comprehensive Model for 2-D and 3-D Engine Simulations, SAE paper 850554.

Blumberg, P. and Kummer, J.T. (1971) "Prediction of NO Formation in Spark-Ignited Engines-An Analysis of Methods of Control," Comb. Sci. Technol. Vol. 4, No. 2, pp. 73-95.

Fenimore, C.P., (1971) "The Formation of Nitric Oxide in Premixed Hydrocarbon Flames", Thirteenth Symposium, The Combustion Institute, Pittsburgh, PA.

Flanagan, P. and Gretsinger, K. (1992) "Factors Influencing Low Emissions Combustion," Proc. of Energy-Source Technology Conference and Exhibition, Houston, Texas, January 26-30. ASME Publication # PD-Vol. 39, Fossil Fuel Combustion, pp. 13-22

Gosman, A.D., (1985) "Aspects of the Simulation of Combustion in Reciprocating Engines," Numerical Simulation of Combustion Phenomena, eds. R. Glowinski, B. Larrouturou, and R.Temam Ispronger-Verlag, Berlin), pp. 46.

Heywood, J.B. (1987) "Fluid Motion Within the Cylinder of Internal Combustion Engines-The 1986 Freeman Scholar Lecture", J. Fluids Engineering, Vol. 109, pp. 3-35.

Kamimoto, T. and Kobayashi, H. (1991) "Combustion Process in Diesel Engines," Prog. Energy and Combustion Science, Vol. 17 pp. 163-189.

Lewis, J.M. and W.T. Tierney, (1981) "United Parcel Service Applies Texaco Stratified Charge Engine Technology to Power Parcel Delivery Vans-Progress Report", SAE paper 801429.

Lewis, J.M. and T.K. McBride, (1983) "UPS Multifuel Stratified Charge Engine Development - Progress Report, SAE paper 831782.

Lewis, J.M., (1986) "UPS Multifuel Stratified Charge Engine Development Program - Field Test", SAE paper 860067.

O'Rourke, P and A.A. Amsden, (1987), "Three Dimensional Numerical Simulations of the UPS-292 Stratified Charge Engine", SAE paper 870597.

Ramos, J.I. (1989) Internal Combustion Engine Modeling, Hemisphere Publ. Corp., New York.

Vargaftik, N.B., (1975) Table on the Thermophysical Properties, John Wiley and Sons Inc.

Westbrook, C. and F. Dryer (1984)"Chemical Kinetic Modeling of Hydrocarbon Combustion), Prog. Energy Combust. Sci., 1984, Vol.10, pp. 1-57.

Zeleznik, F.J. and McBride, B.J. (1985) "Modeling the INternal Combustion Engine," NASA Reference PUblication 1094, Lewis Research Center, Cleveland, Ohio.

Zeleznik, F.J. (1976) "Combustion Modeling in Internal Combustion Engines" Combustion Science and Technology, Vol. 12, pp. 159-164.

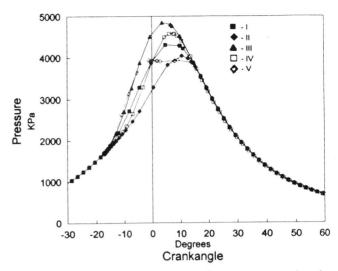

Figure 2. Pressure histories of the n-Octane case, varying the burning formula coefficients in ZMOTTO program.

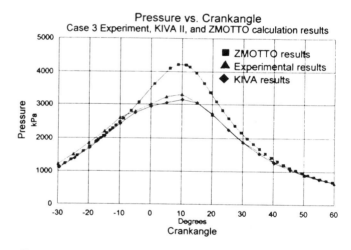

Figure 1. Pressure variation for Case 3; n-Octane, Air/Fuel=61.

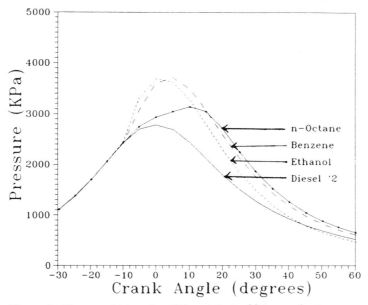

Figure 3. Pressure history for different fuels: Values refer to a moving point at the center of the combustion chamber.

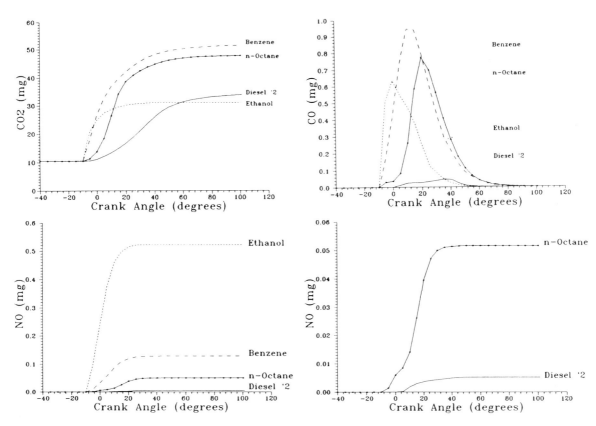

Figure 4. Total species mass production histories for different fuels.

ICE-Vol.18, New Developments in Off-Highway Engines
ASME 1992

RETROFIT OF HIGH TECHNOLOGY DIGITAL IGNITION AND AIR/FUEL RATIO CONTROL SYSTEMS TO MEDIUM AND HIGH SPEED STATIONARY STOICHIOMETRIC ENGINES WITH THE GOALS OF EMISSIONS AND PERFORMANCE IMPROVEMENTS

Richard P. Schook
Altronic, Incorporated
Girard, Ohio

ABSTRACT

Early results from experimental field retrofit to stationary stoichiometric gas engines of a combination system of crankshaft referenced digital ignition and microprocessor based closed loop air/fuel ratio control are described. Improved engine performance and reduced emissions can result from the combined effects of eliminating ignition timing induced variability in cylinder peak pressure angles, stabilizing air/fuel ratio at a point optimum for operation of catalytic converters, and providing automatic control response to variations in ambient and load conditions. A power generator and voltage regulator, which mates to SAE magneto drive flanges, facilitates applications in areas where power is not conveniently available. The equipment is suitable for application in hostile environments and hazardous classified areas.

INTRODUCTION

The development of new electronic engine accessories, crankshaft referenced digital ignition systems and microprocessor based closed-loop air fuel ratio controls among them, has proceeded rapidly. Several prior papers have dealt with electronic air/fuel ratio controllers utilizing technology current as of their date of publication (Engman, 1983; Tiedema and Wolters, 1990), and one dealt with a combination of electronic air/fuel ratio control and digital ignition as applied to slow and medium speed engines (McClendon and Nampon, 1988). Efficiency improvements of as much as 10% have been reported (McClendon and Nampon, 1989).

Application of new accessory technology to medium and high speed stationary engines, particularly in field locations, has lagged behind application on larger slow speed engines. A combination of factors, such as unavailability of convenient power sources for the controls; less emphasis upon fuel efficiency on smaller horsepower units (especially as compared to larger integral compressors); and relatively little regulatory pressure to reduce emissions from existing medium and high speed field units are possible explanations for this condition.

The Clean Air Act of 1990 has provided regulatory impetus to the application of high technology engine accessories to field engines. Implementation details of The Act, which will be enforced by the states via their State Implementation Plans, are still emerging. Specific enforcement guidelines and actions will be dependent upon location of the engine (attainment or non-attainment area), and area classification (marginal to extreme) within those broad classifications, with further classification as a major source (from 100 tons per year per site to 10 tons per year - depending upon classification) or area source (Ealy and Wilke, 1991). It is clear, however, that many stationary field engines will be impacted by the Act's provisions when fully effective.

Regulatory pressure, coupled with increased economic pressure to reduce cost of operations, will likely result in wider retrofit application of technology such as is described in this paper. Reduced emissions, as well as improved engine efficiency and reduced maintenance, can be expected as by-products of crankshaft referenced digital ignition and air/fuel ratio control retrofit.

OBJECTIVE

This paper attempts to extend the boundaries of previous work cited above by focusing on medium and high speed engines and exploring stabilization and control of two critical variables on these engines (ignition timing angle and air/fuel ratio) in a unified fashion. The latest available electronic technology has been employed.

The development and preliminary field test results of two new systems, the CPU-90 crankshaft referenced digital ignition, and the EPC-100 stoichiometric engine air/fuel ratio control, are described. Impact upon emissions, engine performance and operation is explored. Figure 1 presents an overview of the systems.

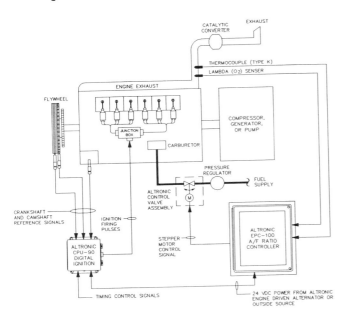

Figure 1
System Overview
(Shown for In-line Engine or
One Bank of V Engine)

DISCUSSION OF CONTROL VARIABLES

Ignition timing angle accuracy and repeatability directly affects the angles at which peak cylinder pressures are developed and therefore, the efficiency of the engine in terms of work performed for a given fuel rate. Some optimum ignition timing angle, which Heywood refers to as maximum brake torque timing (MBT) results in "maximum engine torque ..., maximum brake power and minimum brake specific fuel consumption" (1988, p 828). Variability of ignition timing angle caused by backlash in gears, flexure in chains and couplings, and other mechanical action, causes undesirable deviation from this optimum MBT timing thus reducing engine performance.

Ignition timing angle also affects formation of nitrous oxide, which is related to combustion temperature and duration of combustion. Both of these variables (combustion temperature and time) are affected by the crank angle at which combustion is initiated (Ealy and Wilke, 1991, p. 9). Variability of ignition timing therefore, both cylinder to cylinder and cycle to cycle, detracts from efforts to reduce nitrous oxide formation during the combustion cycle. Uniformity of combustion, and tendency towards detonation are also directly affected by ignition timing angle.

It is not surprising that ignition timing angle has been referred to as "the one, most available, factor that has the most effect upon the economy achieved from an engine ..." (Helmich, 1975). Elimination of timing angle variation, and provision of precise electronic and manual control means for the ignition timing setpoint are natural goals of ignition development (Lepley, 1991).

Air/fuel ratio, as determined by intake air density and temperature, fuel pressure, fuel flow, fuel heat value (BTU) and carburetor adjustment, also directly affects combustion efficiency, fuel rate, power, and emissions. Lambda, the normalized ratio of actual air/fuel ratio to the stoichiometric air/fuel ratio, is a convenient way to refer to the engine air/fuel ratio operating point, and is used throughout this paper. A Lambda less than 1.0 e.g., 0.98 indicates an operating air/fuel ratio slightly rich of stoichiometric, while conversely a Lambda greater than 1.0 indicates an operating air/fuel ratio lean of stoichiometric (excess oxygen).

Air/fuel ratio, along with spark timing, are among "the major operating variables that affect spark-ignition engine performance, efficiency and emissions at any given load and speed ..." (Heywood, 1988, p. 827). It was hypothesized, therefore, that a retrofit combination of crankshaft stabilized digital ignition with advanced timing control features, together with state-of-the-art air/fuel ratio control, could be applied to medium and high speed engines in such a manner as to achieve gains in both emissions and efficiency. Additional benefits of higher reliability were expected, due to solid-state construction and absence of wearing parts in the new systems.

METHOD

A field experiment was designed to test a system based upon the retrofit of new engine accessories designed to stabilize and control ignition timing angle and air/fuel ratio. The equipment retrofit was the Altronic CPU-90 crankshaft referenced digital ignition and the Altronic EPC-100 closed loop microprocessor based air/fuel ratio control.

Retrofit installations were made on three common medium speed stationary stoichiometric natural gas fueled engines: a Caterpillar G399TA (turbocharged and aftercooled), a Superior 8G-825 (naturally aspirated), and a Waukesha L7042G (naturally aspirated). All the engines were in unsheltered field locations, and in continuous operation as prime movers for gas compressors.

To test the effects of the equipment retrofit, and to simulate field variations in ambient and load conditions, attempts were made to collect data in different equipment configurations and under different load and ambient conditions.

Equipment configurations included operation with and without new crankshaft stabilized CPU-90 ignition, and with and without new EPC-100 closed-loop air/fuel ratio control. Observations were made at different ambient conditions (reflecting diurnal variations in temperature) while attempting to hold load and other variables constant.

Attempts were made (within the limits of available field instrumentation) to gather meaningful data which would reflect the following: load conditions, ambient conditions, fuel quality, individual cylinder timing angles, individual cylinder exhaust temperatures, air/fuel ratio (Lambda), exhaust composition, catalytic converter efficiency, indications of fuel rate.

All data was taken with an air/fuel ratio setting slightly rich of stoichiometric, to approximate the optimum operating point of a 3-way catalytic converter (Ballard, 1991, Burns, 1983).

EQUIPMENT DESCRIPTION

An overview of the system under test, which consists of CPU-90 crankshaft referenced digital ignition, and EPC-100 closed loop air/fuel ratio control was shown in Figure 1. A description of the system components follows:

CPU-90 crankshaft referenced digital ignition system:

The CPU-90 Ignition System is digital microcircuit based, and electronically referenced directly to the crankshaft. It has been developed with the goal of minimizing ignition timing angle variation previously caused by electro-mechanical and mechanical phenomena (gears, chains, couplings, etc.). There are no wearing parts in the CPU-90, and the electronics can be mounted off-engine to eliminate vibration effects. It is a low tension design (one ignition coil per cylinder) with electronic distribution of low tension firing pulses to the ignition coils. The capacitor discharge ignition principle is employed. An overview of the system is shown in Figure 2.

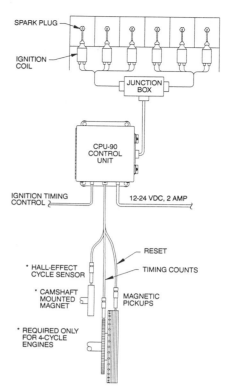

Figure 2
CPU-90 System Diagram

The CPU-90 delivers high energy with a long spark duration, and incorporates switch selectable energy levels. These energy levels can be quantified in terms of millijoules of energy delivered by the primary circuit (which is calculated by taking one-half the product of the primary voltage squared and the storage capacitance in farads) or in terms of the delivered spark duration at a given cylinder voltage demand. This ignition energy, when delivered to the ignition coil (transformer) and stepped up to as much as 40,000 volts, is used first to ionize the gap across the spark plug electrodes, and then to sustain a current flow across the gap for some period of time (spark duration at a given voltage demand).

The two selectable energy levels in the CPU-90 are 90 millijoules (high energy) and 125 millijoules (extra high energy), allowing the user to avoid needless stress on electrical components by operating at the high energy setting if extra high energy is not required. The decision to utilize extra high energy is guided by such factors as presence or absence of secondary shielding, cylinder voltage demand, and engine design ("rich-burn" stoichiometric, open chamber "lean-burn", pre-combustion chamber "lean-burn", pre-stratified charge dilution, etc.). The need for higher ignition energy as a function of a leaner mixture (higher equivalence ratio) has been documented (Lepley, 1991, p. 3).

Five modes of timing control are provided in the CPU-90: multi-position internal switch, single step external switch (internally adjustable increment which is activated by external switch contacts), 1000 ohm external potentiometer, 4 to 20 milliamp current control signal, and internally programmed timing versus engine speed curve. These multiple modes of timing control allow maximum control strategy flexibility.

In operation (see Figure 2), the CPU-90 utilizes a magnetic pickup referenced to the crankshaft ring gear to determine the exact position of the crankshaft in real time. A second magnetic pickup references a ferrous metal reset pin inserted in the crankshaft at the most advanced desired timing of number one cylinder. This reset pickup signals the completion of one crankshaft revolution. Four cycle engines utilize a third pickup, employing the Hall effect (semiconductor sensitive to the presence of a magnetic field), to sense a magnet rotating at camshaft speed. This Hall effect camshaft reference pickup allows the system to distinguish between the compression stroke and exhaust stroke of the engine.

Crankshaft position information is input to digital microcircuits which access firing pattern information stored in a plug-in EPROM memory. The memory customizes the ignition for the particular engine by storing engine and application specific information such as the number of ring gear teeth, timing versus RPM curves (if used), and rate and linearity of timing response to an external 4 to 20 milliamp control signal (if used). All timing adjustments are made in terms of degrees of retard from the pre-determined position of the reset pin, ensuring that timing is never allowed to advance beyond a point which has been determined to be the maximum safe value for the engine.

The CPU-90 is housed in a gasketed NEMA-4 weather-resistant enclosure, and provided with shock mounts to isolate it from engine and foundation vibration.

The resultant ignition system eliminates virtually all timing angle variation and provides extremely accurate firing commands to the low tension distribution circuit. Variations in cylinder peak firing pressures previously caused by ignition timing angle fluctuations are greatly reduced. Since the system utilizes continuous real-time angular reference to the crankshaft, inaccuracies due to misfire induced instantaneous deceleration of the crankshaft are also eliminated.

EPC-100 microprocessor based closed loop air/fuel ratio controller:

The EPC-100 air/fuel ratio control system utilizes an electronically variable Fuel Control Valve(s) installed in series with the carburetor(s) to modulate the fuel flow to the carburetor(s) and maintain the desired exhaust condition (which is an indicator of actual engine Lambda). Closed loop control is utilized. The EPC-100 Controller and associated Fuel Control Valve(s) are a part of the closed loop, as are the carburetor(s) and engine itself. The error signal between the desired and indicated Lambda is evaluated and processed by the EPC-100 software algorithms, then used to drive the loop to a point where minimum error exists.

Development of the EPC-100 has been an attempt to advance the present state of the art in closed loop air/fuel ratio controllers in terms of precision, versatility, and display capability. Design objectives included the following:

- To provide plain language alphanumeric indication of operation and diagnostic information which facilitates both operation of the controller and maintenance of the engine.

- To precisely control the operating air/fuel ratio of the engine so as to maximize the efficiency of a 3-way catalyst.

- To provide maximum stability of air/fuel ratio to avoid loss of load carrying capability despite variations in operating conditions, ambient temperature, and fuel quality.

- To provide a flexible and verifiable means of adjusting control set-point and default actions (the user can manipulate parameters from the sealed membrane keypad and verify the settings on the alphanumeric LCD display).

- To provide a system that is easy to understand, install, and troubleshoot.

In operation (see figure 3) the EPC-100 obtains exhaust information from an inexpensive Zirconia type automotive Lambda sensor which can be used effectively to control near stoichiometry (0.95 to 1.05 Lambda).

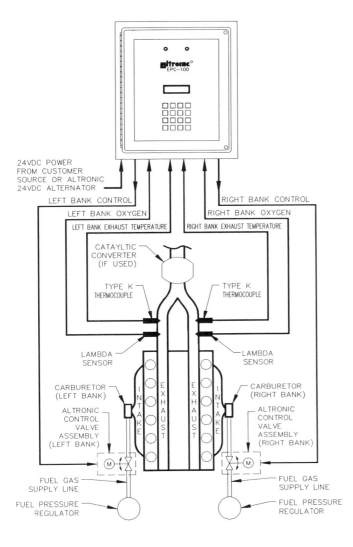

24VDC POWER
FROM CUSTOMER
SOURCE OR ALTRONIC
24VDC ALTERNATOR

Figure 3
EPC-100 System Diagram

The Zirconia Lambda sensor contains a Zirconia element which produces a voltage potential when the partial pressure of oxygen on one side of the element is very small relative to the other side. The element has a catalyzing outer layer which promotes complete combustion of oxygen prior to reaching the Zirconia element. The sensor has very high sensitivity at or near stoichiometry, effectively acting as a switch which provides a low signal when lean of Lambda 1.0, and a high signal when rich of Lambda 1.0. A K-type thermocouple, chosen for its suitability for high temperature operation, is installed adjacent to each Lambda sensor to assure that the operating temperature requirements of the sensor are satisfied (above 343 degrees C/650 degrees F) before closed loop control is activated.

Signal processing and validity testing is utilized within the EPC-100 electronics and software algorithms to take advantage of these sensor characteristics, and also allow a diagnostic to be displayed if the sensor operation appears abnormal. The control algorithms incorporate sensor ready testing, temperature testing, and input voltage testing. In the event of a fault, the EPC-100's alarm output is activated, a diagnostic message is displayed, and manual operation can be selected. The engine can continue to operate in manual mode under operator supervision.

Based upon the validated input of the exhaust Lambda sensor, the Fuel Control Valve's stepper motor controlled actuator is adjusted to maintain the returned sensor voltage at a user defined setpoint voltage. The optimum setpoint is determined by using an exhaust analyzer to find the setpoint for maximum catalyst efficiency (lowest emissions).

Extensive development work has been done to allow the EPC-100 to maintain a setpoint on the steeply non-linear output curve of the Lambda sensor. At the control point, only very small trim adjustments are required. To minimize overshoot, the Controller evaluates the amount of time an error has existed in the feedback loop, and makes adjustments in increments related to the duration of the error signal. The gain of the control is adjustable from the keypad within limits designed so as not to create unstable operation.

A brushless stepper motor, with a permanent magnet armature and a four coil multi-pole stator, is housed within a stainless steel Fuel Control Valve Assembly and installed in series with the fuel line between the fuel pressure regulator(s) and the carburetor(s). The stepper motor has 1,700 steps, and is driven by digital signals from the EPC-100 Control Unit, providing precise position control of a flow restricting piston inside the control valve. This valve, which provides a variable flow restriction between the fuel pressure regulator and the carburetor, provides an air/fuel adjustment function very similar to that traditionally accomplished by the power screw on the carburetor(s).

The Fuel Control Valve actuator position is updated every two seconds, based upon the processed error signal. Actuator position is determined during startup by means of a zeroing and pre-positioning routine. During operation the actuator position is available for operator display, and is an indicator of normal operation as well as bank to bank balance on Vee engines. Lean limit and rich limit position testing is provided, to allow the actuator position to be kept in the center of its control range during setup and present a diagnostic message and alarm if a control limit is reached in automatic operation.

The Fuel Control Valve is non-arcing and non-incendive, as is the stepper motor and EPC-100 Controller itself.

The EPC-100 is configurable to control either an in-line engine with one carburetor or a dual carburetor Vee engine. The Controller incorporates electronic circuitry for either type installation. Most dual carburetor installations utilize a second set of the engine mounted hardware (Lambda sensor, K-type thermocouple, and Fuel Control Valve).

A sealed membrane keypad and a two-line alphanumeric LCD display on the front of the EPC-100 provide a versatile user interface which permits the following functions:

- Viewing of Parameters:
  O2 Sensor Voltage
  O2 Sensor Setpoint Voltage
  Stepper Valve Position
  Stepper Default Position
  Exhaust Temperature
  Estimated Lambda

- Setup of Control Parameters from the Key-Pad
  Setpoint Voltage of O2 Sensor
  Default Motor Position
  Control Gain (response rate)

- Manual Key-Pad Adjustment of Stepper Valve Positions
  Left or Right Bank
  Fast and Slow
  Rich and Lean

- Alphanumeric Diagnostic Fault Descriptions
  Rich or Lean Control Limit
  Exhaust Temp Hi and Low
  Sensor NOT READY Warning

The EPC-100 utilizes a user-defined default stepper motor position to provide repeatable starting conditions and user-specified operating conditions when faults are detected. When the minimum enabling exhaust temperature has been attained, generally within two minutes of loaded operation, the Controller automatically adjusts fuel flow to each carburetor to maintain the desired sensor voltage as specified by the user-defined setpoint. Manual override of the fuel valve position is available for both control channels to aid in setup and troubleshooting.

The EPC-100 also provides two solid state switch outputs (discretes), one of which is used to signal a loss of normal air/fuel control. This switch is software controlled, and indicates an unreliable sensor, lean or rich limit of the Fuel Control Valve, or exhaust temperature above or below acceptable levels. This output could be used to force an engine shutdown via the control panel or to sound an alarm. A display diagnostic message accompanies the switch actuation.

Engine Driven DC Power Source:

The CPU-90 Ignition and EPC-100 Air/fuel ratio control together consume approximately 50 watts of power at 24 volts DC. Should this power not be conveniently available (which is often the case with field compressors), a companion engine driven power source has been developed. This system (see figure 4) mates to standard SAE magneto drive flanges. It utilizes a compact twelve pole permanent magnet alternator to produce AC current. An electronic voltage converter/regulator assembly (mounted off engine), rectifies and regulates the alternator output to produce up to 100 watts (depending upon alternator shaft speed) of 24 volt DC power. Power in excess of that required for operation of the CPU-90 and EPC-100 is available to power control panels, SCADA systems, and similar auxiliary devices.

The DC Alternator system, and its associated voltage converter/regulator, is non-incendive and non-arcing.

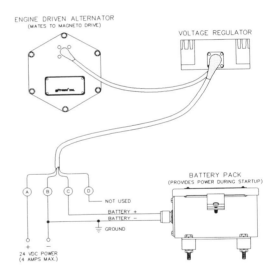

Figure 4
Engine Driven DC Alternator
System Diagram

A diagram of power output versus alternator shaft speed is presented in Figure 5. As shown in the shaded area of the graph, the battery pack provides power during starting conditions, when the alternator shaft speed is below 1000 rpm. In normal operation, the battery pack is charged by the alternator output.

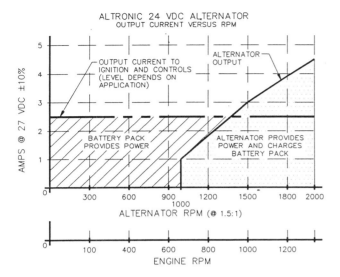

Figure 5
DC Alternator System
Output Characteristics

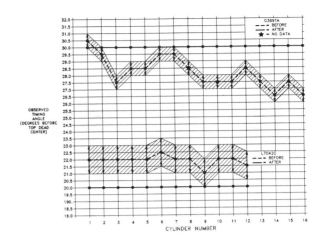

Figure 6
Ignition Timing Angle Observations
Before and After Retrofit
of CPU-90 Ignition

FIELD RESULTS

Field data gathering, and the progress of the field experiment, was hampered by operational restrictions, limited availability of the test engines, limited availability and accuracy of instrumentation at the test sites, and the other assorted difficulties which accompany field testing in an operational "gas patch" environment. Accurate measurements of fuel consumption were not available.

Several observations can, however, be made from the limited data base and accompanying field observations:

1. Ignition timing angle variability has been dramatically reduced, from as much as four degrees cylinder to cylinder to effectively zero. Figure 6 summarizes the timing angle observations before and after change from the existing magneto drive referenced ignition system to the crankshaft referenced CPU-90 digital ignition. Data was available from only two of the three test sites, since installation of the CPU-90 was delayed at the third site (Superior 8G-825).

Cycle to cycle variation on the same cylinder in the before cases is shown by the arrows passing through the average reading (shaded areas on graph). Cylinder to cylinder variation is shown by the overall trend line.

There was no observable variation in timing in the "after CPU-90" case, in either the cycle to cycle or cylinder to cylinder measurements. Engine analyzers (crankshaft encoded) and timing lights were employed as measuring devices.

2. Emissions, as measured at the output of the catalytic converter, were greatly reduced with the use of the EPC-100 in the case where the engine was subjected to a load and speed swing with and without closed loop control. In the uncontrolled engine, cycled from full load to light load, catalyst efficiency decreased 51.5% on CO and 2% on NOX. Corresponding numbers for the same load variation with the Controller closed loop active were an efficiency decrease of 8.1% on CO and an efficiency increase of 1.6% on NOX.

3. A "bonus" of information which facilitated improved maintenance diagnostics and engine performance beyond the action of the control itself was observed. The availability of precise digital readout of Lambda sensor operating voltage, Fuel Control Valve actuator position, and sensor operating temperature, quickly revealed possibly long standing operational problems (such as a large bank to bank imbalance caused by a defective pressure regulator). Corrections were easily made once these parameters were available for routine observation on the EPC-100 display.

4. Individual cylinder exhaust temperatures were slightly improved by retrofit of the CPU-90, with reductions of 3 degrees in bank to bank variation, and 4 degrees in overall deviation on the one engine from which individual cylinder temperature data was available for the

before and after cases (G399TA). The measurements were taken after correction of a pressure regulator defect which was causing a 70 degree bank to bank variation in the "as found" condition (see item 3 above). Other than these reductions, the pattern of cylinder exhaust temperature deviation in the before and after cases appeared very similar and repeatable, indicating that factors other than timing variation were at work.

SUMMARY AND CONCLUSIONS

While accurate quantitative data is difficult to gather from remote field sites, it is clear that there are definite benefits associated with the retrofit of high technology crankshaft referenced digital ignition and microprocessor based closed loop air/fuel ratio control to medium and high speed field engines.

Accurate closed loop control is particularly beneficial on engines subjected to load and speed variations (typical of field gas compressors). Based upon previously cited laboratory work by Heywood and others, it is fair to assume that the elimination of timing variation will result in improved engine performance as well as more effective emissions control.

ACKNOWLEDGEMENTS

Special thanks are extended to the operating companies who graciously volunteered their engines for the equipment retrofit; allowed testing to be integrated with their operations; and provided extensive assistance by their personnel. Altronic Distributor personnel at Hubbell Systems, Inc. particularly Mr. Jeff Kidder, and Ignition Systems and Controls, who arranged and assisted in installation and data collection are also greatly appreciated. Technical assistance, advice, and field efforts from Altronic and Altronic Controls personnel, notably Mr. Gary Kleinfelder, Mr. Gene McClendon, Mr. Maurice Reagan and Mr. Lanny Slater were essential to the project.

REFERENCES

Ballard, Harold, 1991, "Regulations and Emissions Control Technology for Stationary Natural Gas Engines," presented at Pipeline & Compressor Research Council Reciprocating Machinery Conference, Salt Lake City Utah, September 23, 1991.

Burns, K.R., Collins, M.F., and Heck, R.M., 1983, "Catalytic Control of NOX Emissions From Stationary Rich-burning Natural Gas Engines," ASME Paper 83-DGP-12.

Ealy, J. P., 1990, "Reducing NOX Exhaust Emissions On Existing Spark Ignited Integral Gas Engines," Presented at Power Machinery and Compression Conference, Houston, Texas, March 20, 1990.

Ealy, J.P., and Wilke, C. F., 1990, "The Environment and exhaust Emissions Control Techniques," with Appendix I, "Clean Air Act - Acronyms and Key Definitions," presented at Pipeline & Compressor Research Council Reciprocating Machinery Conference, Salt Lake City Utah, September 23, 1991.

Engman, Thomas J., 1983, "Stationary, Gaseous-Fueled, Internal Combustion Engine, Air-Fuel Ratio Control For Application of Three-way Catalysts For Exhaust Emission Reduction," ASME Paper 83-DGP-4.

Helmich, M.J. and Hartwick, W.F., 1973, "Possible Efficiency Improvements Related to Integral Angle Compressors," presented at AGA Transmission Conference, Bal Harbor, Fla., May, 1975

Heywood, Dr. J.B., 1988, _Internal Combustion Engine Fundamentals_, McGraw-Hill, Inc., 1988.

Lepley, J.M., 1991, "Impact of Emission Regulations Upon Ignition System Design," presented at Stationary and Heavy Duty Vehicle Gas Engine Symposium, Didam, Netherlands, April 9, 1991.

McClendon, G and Nampon, C., 1989, "Retrofit Application of Electronic Performance Controls and Digital Ignition to Large Bore, Slow and Medium Speed Engines," ASME Paper, ICE Vol. 9, Book No. 100295, 1989.

Tiedema, P. and Wolters, L., 1990, "Recent Developments in Gas/air Mixers and in Microprocessor Air/fuel Ratio Control Systems."

ICE-Vol.18, New Developments in Off-Highway Engines
ASME 1992

MODERATE COST ELECTRONIC GOVERNING
FOR SMALL OFF-ROAD ENGINES

Hal G. Meyer
Precision Governors, Incorporated
Rockford, Illinois

ABSTRACT

Electronic governing is generally to be desired in all but the most elementary applications. The zero droop capability provides more through-put; the versatility of electronics makes available numerous options and better installations; and the reliability of modern electronics far exceeds mechanical devices. Traditionally, however, electronic governing has been much more expensive than mechanical governing. A series of governors has been designed and produced which brings the price of electronic governors in line with mechanical units.

This paper discusses a number of the design criteria and the logic used in arriving at the present actuator and controller design. Among these are Rotary vs. Linear actuators, Hi frequency vs. Lo frequency speed input, and power limitations due to heat dissipation. Specific advantages of electronic governing for small off-road engines are discussed, particularly for spark-ignited engines.

INTRODUCTION

The design of this series of precision governors was begun in 1978 on the premise that companies would switch to electronic governing if the **reliability** was good, the **application** of the device was simple, and the **price** was moderate. While these goals are easy to state, they have proven far more difficult to accomplish. It is the purpose of this paper to start with these stated goals, and to proceed through the economic and technical evolution which has resulted in the family of control products offered by Precision Governors today. It is expected that this should be of interest on two levels. First, as a view of the development cycle of a product familiar to most people in the engine field, and secondly, as it provides insight into the important aspects that are involved in engine governing. Through a better understanding of this second issue, the engineer or manager should be better able to intelligently evaluate the various options available in the governing field.

RELIABILITY

Let us first consider the matter of **reliability**, and the various factors that produce it. We can generally break the governor into two sections: the **electric actuator**, which provides the muscle to move the fuel system, and the **electronic controller**, which does the computing.

Actuator

The preponderance of **actuators for electronic governing** are of the **magnetic solenoid** type. Gear-motors are rarely used for two reasons: (1) it is hard to make them fast enough, and (2) upon loss of power, they generally fail where they are, rather than being fail-safe; that is, return the linkage to the low fuel position. Magnetic actuators can be further divided into **linear** and **rotary** devices. Both types are currently in use in governor systems. While the linear type is somewhat easier to design initially, in practice it has several shortcomings that are not easy to overcome:

1) Since friction is extremely detrimental to good governing, good anti-friction bearings such as ball bearings or ball bushings are normally required. Ball bushings, commonly found in linear devices, have no equivalent to an inner ball bearing race, and therefore tend to score the shaft on which they ride. In addition to creating roughness, this can also create metal particles, which contaminate the bearing, and lead to failure.

2) The rotor of a rotary actuator can generally be suspended by its bearings such that vibration or shock does not produce unwanted output motion. A linear actuator is far more susceptible to this problem. Vibration or shock loads in a direction which causes motion of the actuator slug will be transferred to the fuel system, resulting in instability or unexplained offspeeds.

3) A linear device with a linear output must generally be mounted in a manner such that the output shaft aims at the fuel system. This is often a complication to a proper and simple mounting bracket. By

contrast, a rotary device can be mounted with considerably more latitude by simply reclamping the output arm on its shaft. While this is done internally in some linear devices, it involves a linkage not needed in a purely rotary unit.

As a result of the above factors, it was decided to use a rotary actuator. Subsequent field experience has validated this decision.

A second point of actuator design philosophy concerns use of a **feedback device** in order to track actuator position. Since the ultimate feedback path is engine speed itself, and this is, of course, continuously monitored, it is only during large load transients that actuator position feedback becomes of significant importance. It is a common practice to substitute the current draw of the actuator as a less expensive analog for actuator position. This is the approach we have taken in the design of this series of precision governors.

The devices available which might be used to provide more accurate feedback include potentiometers, LVDT's, RVDT's, encoders, etc. These devices to date have been characterized by being either fragile and short-lived, expensive, or both. Should a suitable device make its appearance, we would explore its possibilities. Until then, we feel that the price to use direct feed-back, both in dollars and in reliability, is too great for the slightly smaller offspeeds and quicker recovery times one would see during near-maximum load excursions. Although instantaneous full loads are sometimes applied as laboratory tests of engine-governor combinations, these are rarely experienced in actual field conditions.

A third basic consideration in the actuator design involved **size**. A specific amount of work can be extracted from a magnetic device by either: (1) using a substantial amount of steel, and a moderate amount of current, or (2) lesser steel, and substantially more current. Option (1) results in a larger device. Option (2) results in more copper, more heat, higher stress on power handling electronic components, and a smaller device. We chose to lean toward the first option, feeling that it gave us higher reliability and less cost.

In general terms, if one decreases the steel in an actuator by 1/3, say, 3 kg to 2 kg, then the amperage must be doubled to provide a similar work output. This doubling of the electrical energy into the device may nearly double the temperature rise of the coil, drastically lowering its life. Other components exposed to higher current will be similarly affected. It should be clear that pursuing small size by means of using high current levels can seriously limit life and reliability.

Controller

The **electronic controller** circuitry is rather conventional. Like the controller of other electronic governors, it is based on a PID (proportional, integral, derivative) system, and contains appropriate fail-safe circuits for loss of power or signal. It should be noted that mechanical governors are inherently proportional-only devices. It is the addition of the integral path which provides the electronic governor with its zero-droop capability, permitting more horsepower extraction from the same engine. Adding the derivative path provides a degree of "anticipation" to the governor, providing quicker response, and smaller offspeeds.

The selection of the individual **components** to make up this circuitry was done with the highest regard for proven reliability. Each model of integrated circuit, transistor, and other critical part was picked on the basis of a long and troublefree record of use. New "state-of-the-art" chips were passed over in favor of chips with a long record of troublefree service in industrial packages. For example, we chose to use a silicon power Darlington output transistor because of its long history of reliability, rather than try the then-new MOSFET technology.

Board layout was done with generous room for components. Ease of correct assembly, and spacing for adequate heat rejection took precedence over a smaller package. Components are placed tightly to the board and conformal coated for protection from vibration. Additional coats of conformal coating provide further protection against moisture and dirt.

Not every initial decision worked out in practice. Originally the electronics were contained within the actuator housing. Access for tuning was via slots in the housing, which were then sealed with cover plates provided. The thinking was that this eliminated customer wiring between actuator and electronics, and simplified the customers' installation job since only one item required mounting. Better reliability, as well as better customer acceptance was anticipated. Early production experience proved otherwise:

1) At-the-actuator tuning frequently put the tuner in dangerous proximity to hot manifolds or fan blades.

2) All too often, cover plates, which were removed to gain access to the adjustments, were never replaced, leaving the moving parts of the actuator at the mercy of the elements, or less merciful yet, engine pressure washing.

3) Having taken considerable care to provide cool running electronics, we had now subjected them to both engine and actuator generated heat.

The electronics were relocated into their own housing after three years of field experience. A simple two-wire, non-polarized connection between actuator and control box has created no problems. System reliability was improved, and the anticipated negative customer reaction to two modules never materialized.

General philosophy

It is our strong point of view that simplicity is the key to achieving reliable operation while maintaining modest cost. Too often, complexity is confused with cleverness. While a sophisticated device such as an electronic governor requires thoughtful design and modern technology, a conservative approach remains warranted.

SIMPLE INSTALLATION
Actuator

It was previously mentioned that a rotary actuator provides simpler installation than a linear device. This point deserves further discussion.

Fig. (1A) depicts a rotary actuator linked to a linear fuel system, such as a diesel fuel rack. Required travel of the rack is noted by "A", and available rotation of the actuator is noted by "B". It is apparent that the two can be easily matched by making the effective length of the actuator arm the proper distance "C".

FIG. (1A)

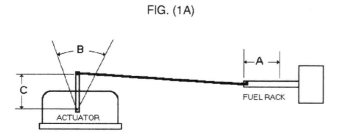

Fig. (1B) depicts the same situation, but with a linear actuator. Now, however, no provision exists to match the two travels when they are dissimilar. If "A" > "B", full rack travel will not be obtained. If "B" > "A", full rack travel will be obtained, but only a portion of the work output of the actuator will be available for use.

FIG. (1B)

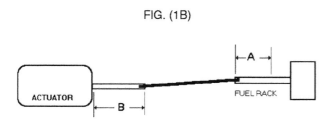

Fig. (2) depicts a rotary actuator linked to a rotary fuel system, such as a carburetor. As in Fig. (1A), it is apparent that selecting a proper effective length actuator arm makes it simple to match the necessary carburetor travel. It should be noted that on many carburetors, the arm is of a fixed effective length, and not easily altered. Thus the same difficulties exist with a linear actuator as in the example in Fig. (1B); the stroke required may not match the travel available.

FIG. (2)

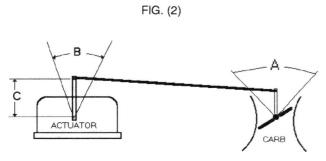

Frequently, there is a convenient place or surface on the engine, but it cannot be used with a linear actuator because the actuator rod must move toward and away from the fuel system. This restriction does not apply with a rotary device with its crank-like arm, capable of being turned on its shaft in order to provide the desired motion. Fig. (3A,B,C) in next column illustrate this point. Further, since the output shaft can extend out both sides of a well designed device, rotating it 180 degrees on its mounting will reverse direction of rotation toward fuel on.

FIG. (3A)

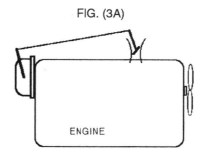

FIG. (3B)

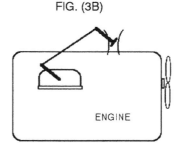

FIG. (3C)

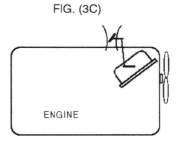

At least one family of linear actuators includes a bellcrank arrangement within the device in order to provide a rotary output. This does, however, add complication to what should be a basically simple device.

Controller

It was presumed from the beginning that **ease of adjustment** was to be preferred over near-infinite tunability. Accordingly, the standard controller has only two adjustments: one to set sensitivity, and one to set speed; and if desired, the speed can be factory set. As previously mentioned, the controller is of the PID type. We use factory established ratios of Proportional to Integral to Derivative, and the normal Sensitivity (or Gain) adjustment scales all three up or down by this same ratio. We have found that this type of adjustability is sufficient for most users. In some cases, a customer requirement will necessitate more or less of one of these functions. An example would be to need more Integral for a gen set application, in order to speed up return to onspeed. In this case, the desired value can be factory set.

The downside of this approach is that one controller cannot be set to fit every job. A controller with six or seven adjustments can more likely be adjusted to serve a gen set, as well as a tow tractor, with one model. We feel, however, that the chosen approach fits the segment of the industry we serve.

This idea of simplicity has been extended to the added functions our governor performs. For example, on spark ignited engines a simple switch in either wire between the controller and the actuator serves as a run-idle selector. Depriving the actuator of power causes it to fall back against the carburetor idle stop.

Similarly, up to three other speeds can be added within the basic controller. The speeds are then selected by installing a switch or relay to send battery voltage to a terminal for that speed. This type of simple **switch logic** makes installation and troubleshooting straightforward— throw the switch, test for voltage at the terminal— if the speed doesn't change, there's a problem in the box!

Terminal blocks were originally selected for wiring connections, and are still in use today. They are hard to beat for simplicity, cost, and ease of use. (Ours feature potted-in connectors, in order to eliminate a leak path for water.) Terminal blocks eliminate the problem of finding a replacement for half of a tough-to-locate connector. The other side of the coin is that it is not only easy to wire them, it is easy to wire them wrong! This problem is particularly troubling in the field, where an untrained end-user may attempt to correct a problem unrelated to the governor by "fixing" the wiring around. In short order, he has a damaged controller to compound his original problem.

There are also installations where **sealed plug-in connectors** provide the best protection against corrosion in a hostile environment. While our basic unit uses terminal blocks, we offer models of all our controllers with sealed plug-in connectors. Our experience seems to be that OEM's in general are moving toward connectors.

A major design goal was to be able to routinely use spark-ignited engine ignition pulses as a free source of engine speed signals. The traditional approach to speed signals for electronic governors has been to require the installation of a magnetic pickup. This means the purchase of a fairly expensive pickup, and then drilling, tapping and installing the pickup in the flywheel housing. This may make sense for diesel or large gas engines, but in the small gasoline engine market, it adds a degree of expense and complication that is desirable to avoid if possible.

The advantage of a magnetic pickup is that it provides a smooth wave form at a frequency of about 4000 Hz, whereas the engine ignition system is irregular, terribly noisy, and only runs about 100 Hz. We were able to pioneer the use of this ignition signal in 1978, and have refined such techniques since, thereby making the additional complication of a magnetic pickup unwarranted and unnecessary in most applications. When a magnetic pickup is used with our controller, as in a diesel application, we merely include an IC which divides the frequency by 40, and thereafter treat it as we would an ignition signal. The disadvantage here is that speed information is not updated as rapidly. In practice, this is only detectable during full load changes.

General

Simple test procedures provided in the installation manual permit even a modestly qualified user to determine which, if either, part of the governor system is at fault. The low cost of half a system (an actuator or a controller) makes replacement a very reasonable alternative to on-site trouble-shooting. Repairing such units generally runs 40-50% of their price, and are routinely accomplished in a week. **Simplicity of service work** is thus accomplished as a result of modest cost.

PRICING

Any successful attempt to displace old technology with new must do so not only with a superior product, but also without inducing "sticker-shock". Accordingly, we established price goals that would be reasonable when compared with mechanical governors, after considering the inherent advantages of reliability, versatility, and better engine performance. Typically, a **basic governor** at small monthly OEM quantities, with a run-idle capability, zero-droop, terminal block connectors, and an actuator sized for a four cylinder engine currently sells for **$125**. Whistles and bells (multiple speed set points, internal overspeed shut-down, magnetic pickup option for diesels, distributor-less ignition engine dual signal input, sealed connectors) add 5 - 15% to the cost.

FOR THE FUTURE

The next product for this precision governor series is now in the final stages of R & D, and should be ready for extensive field testing in about six months. This new rotary actuator has a useful travel of 65 degrees, far beyond that available on current actuators, typically 15-36 degrees. This, in conjunction with innovative mounting and coupling techniques, will make possible direct attachment to the carburetor, and eliminate linkage, and attendant difficulties. It is sized for four cylinder gas/gasoline engines. In addition, it appears that cost control through engineering design decisions which are highly biased by cost considerations will permit selling prices substantially below our current governors.

SUMMARY

We have been able to successfully design an electronic governor system suitable for small diesel and spark ignited engines in off-road applications, which sell for a moderate price. Its actuator is of rotary design, while the electronics is housed in a separate package. Its speed signal can be provided by a magnetic pickup on a diesel, or from the ignition system on a spark engine. It is intended to be simple in design, construction, and in use. It can provide an affordable option to mechanical governing, bringing a host of advantages at modest additional cost.

All engineering (except very bad engineering) involves compromise. We have attempted to detail in this paper some of the choices we were forced to make in order to achieve the final product. Some of these decisions resulted in lowered accuracy and dynamic performance, compared to a governor system costing two to ten times as much. We accept these limitations, and contend that a large number of applications exist where they are of little concern compared to the overall improvement our system offers over mechanical governors.

ADDENDUM:
Electronic Governing Fundamentals

The following discussion should provide a general understanding of how electronic governors function. It is for persons new to this field, and is intended only to provide background rather than detailed instruction. The information to follow pertains to most electronic governing systems, but not necessarily all.

Controller

The **controller** accepts a train of pulses from the engine (via the ignition system, magnetic pickup, or other sensor). It takes note of the time between pulses, and from this deduces the RPM of the engine. This **actual engine speed** is converted to a DC voltage level. This voltage is compared to another DC voltage which represents the **desired engine speed,** and is usually set by means of a user-adjusted potentiometer. The difference between these two signals, if any, is a third voltage, known as the **error signal**, which is the basis for all subsequent calculations.

This speed error is used three ways in order to give the best possible form of engine control.

1. **Proportional**—Corrective action is taken proportionally to the magnitude of the speed error. That is, twice the action is taken for an error of 200 RPM as for 100 RPM.

2. **Derivative**—Corrective action is taken on the basis of how fast the error is changing, and whether it is getting better or worse. This method permits the governor to take note of rapid speed changes caused by sudden load variations, and to begin to take action before the speed error becomes large.

3. **Integral**—Corrective action is taken on the basis of the size of the error, **and** on how long the error has existed. Under this method, any speed error, no matter how small, will result in a continually increasing corrective action, until the error is finally reduced to zero. This is the scheme by which isochronous operation, or "zero droop", is achieved.

The relative amount of each of the above corrections can either be determined by the governor manufacturer, or left to the user as an adjustment. In any case, the **corrective action signal** is the sum of the combined efforts of the above three computational systems.

This corrective action signal is next ratioed up or down by a **gain** or sensitivity adjustment. This adjustment sets the response rate of the governor, and is set according to the overall dynamic characteristics of the total system being controlled. Too much gain results in unstable operation; too little will cause sluggish response.

The final stage of the controller is the **power handling** section. Here the small signal from the computations described above controls large amounts of power sent to the actuator. Thus the electrical power for actuator muscle is provided.

Actuator

This device converts the electrical power from the controller into **mechanical force** or torque. A sketch of a generalized rotary actuator is shown in Fig. (4).

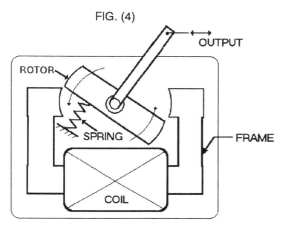

FIG. (4)

As amperage is applied to the coil by the controller, the steel frame and steel rotor act as an electromagnet. The strength of this electromagnet is varied according to the amount of current sent by the controller. This magnetic field acts to pull the rotor into registration with the frame (rotate the rotor into a horizontal position). This magnetic force is opposed by the spring, which wants to push the rotor back the other way. In a properly designed actuator, each additional increment of current increases the magnetic force by a fixed amount, compressing the spring by a fixed amount, resulting in a fixed amount of further rotation of the rotor (and thereby, the output arm). Thus, the output arm position is proportional to the amount of current applied by the controller. In this manner, the fuel system metering device can be precisely positioned according to the calculations of the controller.

Isochronous Control

This major advantage of electronic governing is often not clearly understood. Isochronous merely means that when a load is applied to the engine, the speed will return precisely to the same speed as when unloaded. This can be compared to droop governing, where loaded speed will be 5%, 10%, or more below the unloaded speed.

It is easy to see the advantage in a gen set, where frequency must be maintained, but where is the advantage in, say, an engine for construction equipment? It is due to the fact that **power = torque X RPM,** so if the engine speed can be maintained, more power can be extracted from the same engine. Therefore, an isochronous governor can permit more output from an existing engine, or the substitution of a smaller engine to produce the same output power.

ENGINE POWERED IGINITION, NO WEARING PARTS

Joe B. Stephens
Dynalco Controls
Fort Lauderdale, Florida

ABSTRACT

This paper describes a novel ignition system that derives its energy from a special form of alternator that is attached to the engine crankshaft. The paper shows how the alternator is constructed in such a way as to have no bearing or wearing parts. A magnetic ring assembly that is attached to either the fan belt pulley or the flywheel of the engine is the heart of this generator system. Selection of the type magnetic materials used and how the magnets are placed to obtain maximum benefit from their energy content is also discussed. Also discussed is how the generator windings are placed and how the electronics were designed to make optimum use of those windings. Also explained is the Hall effect trigger system and how the voltage is derived to operate the Hall effect device. Voltage curves that show how both low and high engine speeds are accommodated are explained. A comparison is developed to show how the performance of this new system overcomes the disadvantages of the original equipment systems design. Specific advantages of crankshaft triggering and additional energy delivered to the spark plug will be shown.

Introduction

Magnetos have been used to supply the spark to gas engines where batteries were not practical for many years. Although the magneto did a good job there were many maintenance problems associated with bearings, couplings, gears and breaker points to name a few.

The thrust of this development is to eliminate many of the problems discussed above. The first step was to eliminate as many moving and wearing parts as possible.

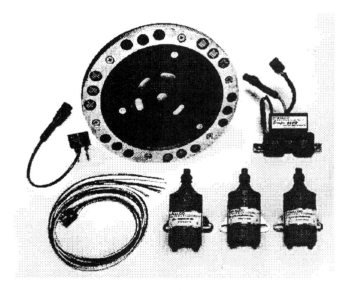

This was accomplished by mounting a steel plate approximately 10" in diameter and 1/4 inch thick to the engine crankshaft, usually on the radiator end of the engine. This plate has an aluminum ring mounted on the outer edge that has Neodymium magnets epoxied into it. Neodymium magnets really made this design easier because of their high magnetic strength and high coercive force. The size of the magnets are 1/2 inch cube and three of these magnets are used per charge and are spaced a pole distance apart, which in this case is one inch center line to center line. There are three sets of three magnets on the plate for a 6 cylinder, 4 cycle engine, and two sets of three magnets on a 4 cylinder, 4 cycle engine, and etc. Three magnets per charge is used to get several flux reversals per charge.

One of the problems encountered in this design was mounting of the steel plate to the engine, but has been over come by ingenuity and a small modification.

Now with the plate mounted to the crankshaft and the engine rotating the magnets will be passing by a generator mounted near the rotating magnets. The generator will be discussed later.

The steel plate that has the generator magnets also has the trigger magnets that determine where the engine will spark. In the case of a 6 cylinder, 4 cycle engine there are three 1/4 x 1/4 inch Neodymium magnets spaced 120° apart on different radii. The engine will fire on the exhaust stroke when this type triggering is used on a 4 cycle engine. If the triggering is derived from a 1/2 engine speed shaft the system does not fire on exhaust, all else works the same.

A small 1/2 by 2 inch trigger box in which the trigger Hall devices are mounted faces off to the trigger magnets mounted in the rotating plate that is attached to the crankshaft. The Hall devices are solid state electronic switches that switch on when exposed to a magnet of proper polarity, and sufficient magnetic strength and will switch off when the magnet has passed. A Hall device is affected little by the velocity at which the magnet passes, therefore no appreciable time shift is seen as velocity changes, also if voltage to operate the device is available it becomes a zero velocity switch which results in low engine speed starting. All this results in a very accurate and reliable low speed timing device.

Generator:

The generator faces off to the magnets located in the rotating plate mounted on the crankshaft. The generator is usually mounted on a bracket that is attached to the engine block or frame. The electronic package is usually attached to the generator, but can be mounted separately if so desired.

The generator is made up of a laminated steel "E" core that contain two windings. One winding is made up of many turns of small wire for the purpose of developing a high voltage at low speeds, but tapers off at higher speed because of the large amount of inductants.

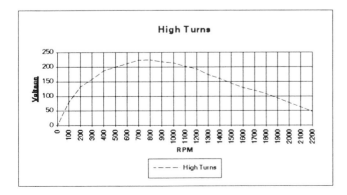

The second winding is made up of fewer turns of a larger wire and produces a high voltage at high speeds.

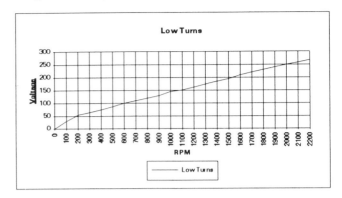

The two windings are connected together so their outputs will be series aiding so the benefit of both winding are realized.

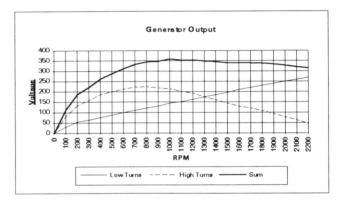

During the starting phase the winding with many turns predominates, but at operating speeds of the engine the winding of fewer turns predominates. The windings are completely encapsulated in epoxy to protect against moisture.

If there is a need for 15-20 Watts of power another generator can be mounted just to produce power for other needs.

The air gap (spacing between generator and magnet) can operate much wider using Neodymium magnets, it is not unusual to operate with .060 to .080 inches air gap on both generator and trigger.

Ignition:

The ignition system is a high energy capacitor discharge type. The operating voltage is 300 volts and the discharge capacitor is 4 micro farad which produces 180 millijoules of energy.

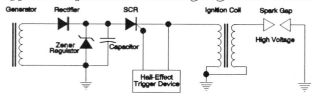

Typical Capacitive Discharge Ignition Circuit

The generator charges the storage capacitor and produces low voltage for the Hall trigger devices. The Hall devices trigger the main S.C.R.'s that discharge the storage capacitor into the ignition coils. The high energy ignition gives good low speed starting and long spark plug life.

All the electronics are epoxy encapsulated to protect against physical abuse and prevent moisture intrusion. The encapsulated electronic parts are modular, and low cost to provide for a quickly replaceable maintenance technique which eliminates the requirement for highly skilled repair personnel thereby reducing maintenance cost

Conclusion:

The objectives of this development project have been met, no wearing parts, long life electronics, high energy ignition and resistance to moisture. There are several hundred engines of 4 different manufacturers running with good results. Although initial installation required new bracketry and fabrication, and some additional training of personnel, the results more than offset these "up front" costs, and yielded a system which met all aspects of the design goods..

INTEGRAL ELECTRIC ACTUATOR
APPLIED TO A DB2 INJECTION PUMP

George Anderson
Barber-Colman Company
Rockford, Illinois

ABSTRACT

An electronic actuator has been developed which is housed in a fuel pump cover. This actuator modulates the pump fuel metering valve to supply the appropriate amount of fuel for engine speed governing.

The pump requires no changes from a standard unit, only the replacement of the cover. The integrity of the in-pump mechanical governor remains intact. An independent engine overspeed can be maintained through the mechanical governor.

The electric actuator is accessed through a cover mounted terminal strip. Several electronic governor packages are available to implement various control algorithms.

The Stanadyne injector pump is used in a wide variety of low horsepower high speed diesel engines. Figure 1 shows a cross-sectional view of the injector pump from the Stanadyne manual. On the far left-hand side of the injector pump is the mechanical coupling to the engine drive, which supplies power to the injector pump and also serves as a speed reference for the mechanical governor. The flyweights for the governor are on the left-hand side just inside the enclosure. Also connected to the flyweight mechanism is the speeder spring which is in turn positioned by the throttle lever. The metering valve is shown as item 11 in the upper right-hand portion of the cross-sectional view. The metering valve is driven through a linkage connected to the mechanical governor. The external throttle setting lever varies the force exerted on the speeder spring which in turn determines the amount of balancing force required from the flyweights to position the fuel metering valve.

Figure 2 shows a standard variable speed droop governor characteristic with the RPM axis ranging from idle speed to rated speed and the droop line being a function of load applied to the engine.

To apply the electronic actuator to the injection pump, the cover shown in Figure 1 of the injection pump is removed. This cover is replaced with a cover that has basically the same footprint area of the standard Stanadyne fuel pump cover which houses the electronic actuator shown in Figure 3. The output mechanism of the actuator is a forked-arm which engages into the fuel pump linkage which controls the fuel metering valve described in Figure 1. The external connection to the electronic actuator is a terminal strip with two terminals which connect to the electronic actuator control coil. The second connection to the actuator cover is the standard output for drain fuel. This is identical to the standard drain fuel outlet which is a part of the standard Stanadyne pump cover.

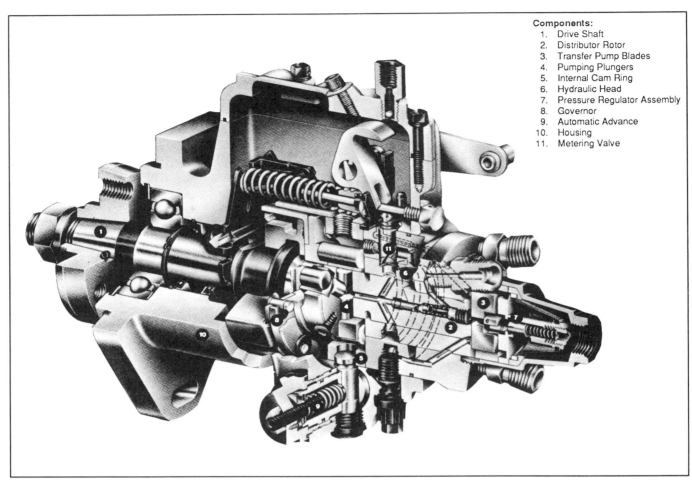

Components:
1. Drive Shaft
2. Distributor Rotor
3. Transfer Pump Blades
4. Pumping Plungers
5. Internal Cam Ring
6. Hydraulic Head
7. Pressure Regulator Assembly
8. Governor
9. Automatic Advance
10. Housing
11. Metering Valve

FIGURE 1
CUT AWAY VIEW OF STANADYNE ROTARY FUEL PUMP

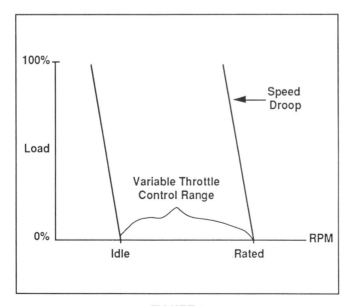

FIGURE 2
VARIABLE SPEED MECHANICAL
GOVERNING FUNCTION IN THE STANADYNE PUMP

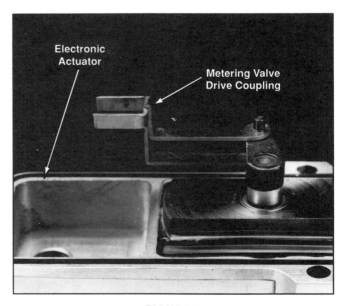

FIGURE 3
ELECTRONIC ACTUATOR
WHICH REPLACES THE STANADYNE COVER

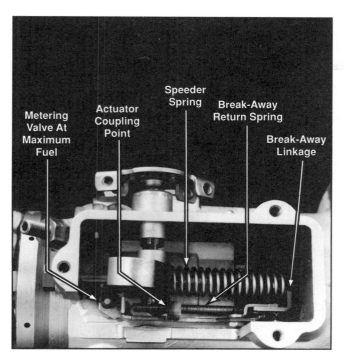

FIGURE 4
CUT AWAY VIEW OF ELECTRONIC ACTUATOR
SHOWING METERING VALVE AT MAXIMUM FUEL

FIGURE 5
CUT AWAY VIEW OF ELECTRONIC ACTUATOR
SHOWING METERING VALVE AT MINIMUM FUEL

Figures 4 and 5 are photographs of a cutaway of an actual Stanadyne rotary fuel pump. Figure 4 shows the mechanical governor linkage in the maximum fuel position. Figure 5 shows the metering valve linkage in the minimum fuel position. Notice in Figures 4 and 5 the speeder spring is compressed, calling for full pump output fuel, and in Figure 5 the metering valve has been positioned to the zero fuel position. This is accomplished through a break-away linkage point within the mechanical governor linkage. The break-away linkage return spring is extended in this position and if it is not held in this position, the return spring would pull the metering valve linkage back to the maximum fuel position.

Figure 6 shows the electronic actuator in a partially installed position with the metering valve linkage partially engaged to the fuel linkage coupling point.

Figure 7 shows a fully installed electronic actuator at the zero power condition due to the electronic control actuator. (The normal position of the metering valve would be at full fuel at engine start-up when the mechanical governor is in control.) But with the electronic actuator installed, the metering valve position is overridden and taken to the zero fuel position. For the moment assume that the mechanical governor setting is at a higher speed than which the electronic actuator can control speed. At engine start-up the electronic governor would sense an underspeed condition and move the metering valve towards the full fuel position. This allows the mechanical governor break-away linkage spring to return to its relaxed position and the electronic governor is then set at full fuel to the engine. Upon engine start-up and approaching the electronic governor set speed, the actuator will extend the break-away return spring. This reduces the amount of fuel to the engine to maintain running RPM at set speed.

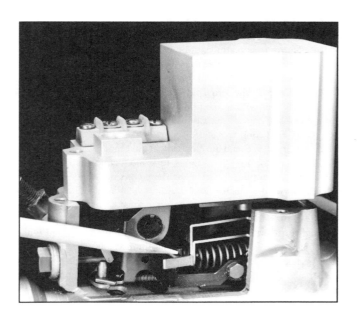

FIGURE 6
PARTIALLY INSTALLED ACTUATOR
(ACTUATOR FORK APPROACHING
THE FUEL METERING LINKAGE)

101

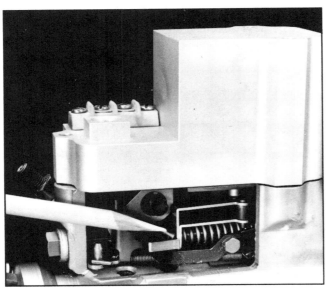

**FIGURE 7
FULLY INSTALLED ACTUATOR
(ACTUATOR FORK FULLY ENGAGED
TO THE FUEL METERING LINKAGE)**

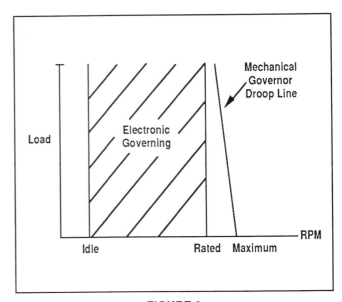

**FIGURE 8
REGION OF RPM CONTROL
AVAILABLE TO ELECTRONIC GOVERNING**

Figure 8 is a graphical representation of the region which the electronic governor can control engine speed. Notice that the mechanical governor speed setting is greater than the speed available to the electronic governor. The electronic governor can control speed all the way from a low idle to the rated electronic governor speed. The electronic governor cannot control speeds greater than that set by the mechanical governor. If the electronic governor attempts to go into this control region, the metering valve linkage will be pulled away from the output of the electronic governor by the mechanical governor linkage.

The electronic actuator coupling linkage consists of a fork arrangement which is designed to insure that engagement will always exist between the mechanical governor linkage and the electronic governor control. The length of the electronic governor coupling fork is such that with maximum travel on the electronic actuator, minimum metering valve position will not cause decoupling of these two points. This condition could occur when the electronic governor is calling for full fuel and the mechanical governor is calling for zero fuel. The actuator linkage coupling fork and the mechanical governor actuator coupling point were key considerations since the installation of the electronic actuator is a blind installation. The installation of the electronic actuator and the coupling of the output linkage of the actuator to the actuator coupling point in the pump is a straightforward process which can be easily demonstrated with actual hardware. The actuator dynamic characteristics were considered by taking into account the break-away return spring and the proper sizing of the actuator return spring. At engine shutdown the actuator power is removed and the spring force within the actuator must return the metering valve to the zero fuel position. The mechanical stroke of the electronic actuator output is sized to assure minimum to maximum positioning of the internal pump metering

valve. The actuator is available in 12 volts, 5.5 amps maximum current with 1.2 to 1.5 amps required in the governing range. A 24 volt version is available, 3.2 amps maximum current with 0.6 to 0.9 amps required in the governing range.

There are distinct application advantages to applying the electronic actuator internal to the fuel pump. One advantage is the removal of all external pump linkages normally required when an electronic governor is added externally to either the fuel shut-off lever or throttle setting lever. This results in a substantial savings in time and labor and external bracketry costs. A second advantage is if an application requires remote speed control of the engine, it can be accomplished easily without cumbersome mechanical linkages to the throttle lever.

ICE-Vol.18, New Developments in Off-Highway Engines
ASME 1992

THE ADVANCED DIESEL ENGINE FOR
OFF-HIGHWAY APPLICATIONS

C. H. Such, D. M. Doyle, and D. Broome
Ricardo Consulting Engineers Limited
Shoreham-by-Sea
West Sussex, United Kingdom

ABSTRACT

The off-highway diesel engine is now faced with increasingly stringent emissions standards. In this paper Ricardo presents and discusses the key technologies that will be used to meet the challenge. Attention is focussed on the trade-off between meeting the emissions targets on the one hand and achieving competitive engine performance at an acceptable cost on the other hand.

1. INTRODUCTION

Automotive direct injection (DI) diesel engines have been the subject of much development in recent years, with the objectives of reducing exhaust emissions, improving fuel consumption and power output. Much of this activity has been concentrated on DI engines used in commercial vehicles, which have been the subject of increasingly severe emissions legislation in many parts of the world.

Recently, off-highway engines used in a wide variety of applications have come under scrutiny regarding their contribution to the total exhaust emissions from internal combustion engines. Regulations on the gaseous and particulate emissions from off-highway engines have been proposed in California and are under discussion in Europe and in Japan.

For both on-highway and off-highway engines, the emissions standards must be achieved without sacrificing the established requirements for power, torque, fuel consumption, reliability, durability and, very importantly, low first cost.

At Ricardo, many years of experience and success in meeting emissions legislation has resulted in a systematic approach to the problem of combustion system development for the DI engine [1,2,3,4]*. Due to the complex nature of diesel combustion processes, many aspects of development are still the subject of trial and error, with every engine requiring individual attention. However a number of basic principles have emerged which allow the development effort to be minimised and a better balance between the emissions, economy and performance objectives to be obtained.

Based on Ricardo's recent experience, this paper reviews some of the key features of low emission combustion systems which can be applied to off-highway DI engines of between 0.8 litre and 2 litre /cylinder. Reference is also made to the potential of the indirect injection engine of between 0.4 litre and 0.9 litre/cylinder.

2. EXHAUST EMISSION REGULATIONS

2.1 Current Status

The situation regarding off-highway exhaust emission regulations is still confused at the time of writing. Proposals have been made by CARB and it is likely that these limits - shown in Table 1 - will be introduced from 1996 for engines over 175 bhp in California. The US EPA is currently considering suitable regulations for smaller engines.

The basis of CARB's proposed standards is that the off-highway engines should meet limits equivalent to those met by 1990 on-highway engines, starting in 1996. From the year 2001, they should meet limits equivalent to 1991 on-highway standards.

In Europe, exhaust emissions standards do not yet apply to off-highway engines. Proposals have recently been made by EUROMOT, the group of European-based engine manufacturers and these are now being considered by the European Commission. The basis of the proposals is that smaller, less powerful, engines are allowed higher specific emission limits, reflecting the fact that they need to be relatively cheap and can not accept the cost increase of extra low emissions technology. The limits therefore reduce with increasing power up to 150 kW and remain constant above 150 kW.

In Japan, it is understood that exhaust emissions limits will apply to engines used in tunnels on government supported construction

*Numbers in brackets designate the references at the back of the paper.

TABLE 1
CARB PROPOSED OFF-HIGHWAY STANDARDS
8 MODE TEST

	1996	2001
HC g/bhp.h	1.0	1.0
NOx g/bhp.h	6.9	5.8
CO g/bhp.h	8.5	8.5
Particulates g/bhp.h	0.4	0.16
Smoke opacity		
Lug %	15	15
Accel %	20	20
Peak %	50	35

Notes:
- 175-750 bhp Diesels only
- Closed crankcase breather for NA engines only
- No durability test
- Quality audit test: minimum of 3 engines or 1% or sales taken from end of production line
- Low sulphur, low aromatic fuel allowed for certification

projects from 1996, and in other applications one year later. The CARB Stage 1 (1996) limits are likely to be adopted.

2.2 Test Cycle

The cycle selected by CARB for off-highway engines is the 8-mode cycle, otherwise defined as cycle C1 according to the ISO Draft Proposal 8178. This cycle is intended to be representative of off-highway engines running at medium to high load in construction, agricultural and forestry equipment. The modes and weighting factors are as follows:

Speed	Torque %	Weighting
Idle	-	.15
Max torque	50	.10
"	75	.10
"	100	.10
Rated	100	.15
"	75	.15
"	50	.15
"	10	.10

The EUROMOT proposal also adopts the cycle C1 as a reference, although alternative cycles and different emissions limits may be applied to engines used in different applications. The background to this was discussed in Reference [5].

It is likely that cycle C1 will also be adopted in Japan for off-highway engines, making the cycle perhaps the only example of a globally-accepted emission test cycle.

Unlike the various on-highway test cycles, the cycle C1 consists of 8 modes each with a relatively even weighting factor, which means that it is important for the emissions to be controlled equally well over most of the operating map.

3. COMBUSTION SYSTEM DESIGN FOR LOW EMISSIONS

The DI engine type is used almost universally in off-highway engines above about 50 kW because of its excellent fuel consumption and good starting characteristics. The following sections refer to design principles for multi-cylinder DI engines of about 0.8 litre - 2 litre/cylinder.

3.1 Criteria for Good Combustion

Fuel consumption is only one of the indicators of the quality of the combustion process. Low values of fuel consumption can be achieved in DI engines of conventional design by the use of advanced combustion timings, however this is usually accompanied by unacceptably high NOx emissions, noise and cylinder pressures. A more realistic development objective for modern DI engines is the achievement of minimum fuel consumption at relatively retarded combustion timings. To meet future NOx standards, combustion at some operating conditions must be retarded beyond the point of minimum fuel consumption. It is therefore vital that NOx emissions at a given start of combustion are minimised by correct design of the combustion system, to avoid excessive timing retard and hence BSFC penalties.

A second important consideration is the achievement of minimum emission of carbon (ie smoke). Carbon is of importance because it comprises a large part of the exhaust particulate and in fact represents the major development challenge in particulate reduction. Typical particulate compositions from a naturally aspirated DI engine are shown in Table 2. The carbon fraction is the result of two main sources, operation at richer air-fuel ratios (AFR) at high load, and from incomplete combustion of fuel at part load conditions. Although the dramatic increase of smoke emissions which occurs towards limiting AFR can be moderated by the use of high pressure fuel injection equipment, part load carbon is more difficult to control and is present even when conventional smoke readings indicate zero. For emissions cycles with significant part load weightings, such as the 8-mode cycle, it is essential to achieve low values of part load carbon emissions.

3.2 The Importance of the Injection System

The key to achieving low carbon emission is the fuel injection system. The ideal injection system has the following main features:

-   High injection pressure ability, which allows small nozzle total area and controlled injection duration to give low carbon formation at full load.
-   Rising injection rate to control NOx.
-   Rapid end of injection ("spill") to control carbon fraction at part load.
-   Flexible control of injection timing.

The DI combustion process depends primarily on the mixing energy available from the injection system and the organised air motion (swirl) presented in the intake charge. Typically, the optimum swirl ratio varies across the engine's operating range, being lower at high speed for open chamber combustion systems, and the magnitude depending on the injection energy [Reference 6,7]. Maximum injection pressure can be taken as a broad measure of injection energy (hence mixing potential), however a more reliable

104

TABLE 2
TYPICAL PARTICULATE COMPOSITION FOR NA DI ENGINE
MAXIMUM INJECTION PRESSURE 600 bar FUEL SULPHUR 0.05%

	Rated Speed Full Load	Rated Speed 50% Load	Rated Speed 10% Load
	(g/bhp.h)		
Carbon	.18	.10	.28
Fuel HC	.03	04	.53
Oil HC	.01	.02	.30
SO + H O	.01	.01	.02
Remainder	.05	.03	.19
Total	.28	.20	1.32
Smoke (Bosch)	2.1	0.6	0.3

indicator used routinely at Ricardo is the mean effective injection pressure (MEIP), representing the mean pressure drop across the nozzle holes calculated from Bernoulli's equation [Reference 8]. This takes into account the injection period, fuelling quantity and the nozzle area and discharge characteristics. Typical MEIP and maximum nozzle pressure characteristics for different injection systems are shown in Figure 1. The latest designs of heavy duty electronic unit injector and sleeve controlled in-line fuel pumps [Reference 9,10,11] possess similar MEIP characteristics across most of the torque curve.

At high load, NOx emissions tend to increase with higher MEIP, whilst particulate tends to reduce as the carbon fraction becomes smaller. DI combustion systems can be designed to take more or less advantage of the injection energy, depending on the engine type, the emissions targets and other constraints such as the pressure capability of the injection system, as described later. With all types of combustion system, high injection pressures at low speed are essential to maintain low smoke emissions and hence high torque. The ability of the sleeve-controlled injection systems to maximise low speed MEIP within a given maximum pressure limitation has contributed to the success of these systems for demonstrating low emissions [Reference 1,2,12]. The latest designs of electronic unit injector can also match these low speed pressure characteristics, albeit with higher maximum injection pressure at full power (Figure 1). The high pressure common rail injection system can also achieve high injection pressure at low speed and has been shown to give very good performance and emission results [Reference 13,14,15].

The particulate formation process is also very dependent on the detailed characteristics of the injection event. High initial rates of injection can give increased premixed combustion, leading to increased NOx and HC emissions and combustion noise. However for modern low emission turbocharged medium and heavy duty DI engines operating with high load factors these effects are minimised due to very short ignition delay. In naturally aspirated (NA) medium duty and light duty engines the fuel injected during the delay period has a more important effect, hence the use of high compression ratios and measures such as 2-spring injectors to shorten ignition delay and control initial injection rate in these engine types [Reference 16].

The quality of the end of injection has a significant effect on particulate emissions. Rapid rates of spill at full load conditions effectively give a higher MEIP. At light load a high degree of fuel

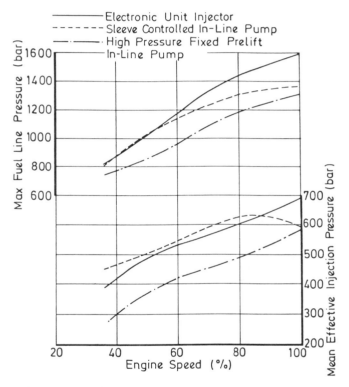

Figure 1   Fuel injection pressure characteristics for heavy duty DI engines (Reference 3)

atomisation must be maintained up to the end of injection, hence rapid spill is important to control formation of carbon (Figure 2). The correlation between spill rate and carbon emission tends to be seen at more advanced timings, becoming less clear at retarded timings due to the onset of combustion instability. Consequently the full benefit of a rapid spill system would be obtained only where flexible timing control could be used to provide light load timing advance. A practical limit exists for rapid spill rate due to the need to limit gas blow-back into the injector nozzle.

Injector nozzle design is of critical importance. The most obvious requirement is for minimised sac volume to reduce HC emissions

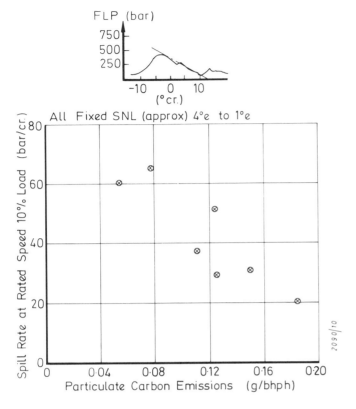

Figure 2   *Effect of pump spill rate on rated speed 10% load particulate carbon emissions at approximately constant timing (Reference 12)*

and particulate volatile organic fraction (VOF). This is especially important for medium duty engines with low ratings (eg naturally aspirated) in which case the volume of the nozzle sac and nozzle holes (typically 0.3-0.5mm³) is a

significant proportion of the total injected fuel volume. The minimum HC emissions would normally be achieved by a valve covering orifice (VCO) nozzle but VCO nozzles can give very different mixing characteristics to minisac types, sometimes affecting the emission of carbon adversely [Reference 17].

3.3 Choice of Combustion System

Two main DI combustion chamber options are available to the engine designer. Open chamber (toroidal) combustion bowls have traditionally been used in most medium and heavy duty DI engines, whilst re-entrant combustion chambers have found favour in smaller light duty engines. The choice primarily depends on the capability of the fuel injection system and the size and application of the engine.

Re-entrant combustion chambers have the principal advantage of operating more effectively with lower injection pressure than their open chamber counterparts. Due to the ability of the chamber to retain a high degree of in-bowl air motion after TDC, optimum smoke and fuel consumption timings tend to be more retarded than for open chambers. Relatively high swirl ratios are needed, typically

between 2.0 and 3.0 Rs (ie measured by impulse swirl meter) to give a good match across the engine operating range.

For naturally aspirated engines of about 1 litre/cylinder, increasing injection pressure results in improving total particulate and carbon fraction at rated power (Figure 3). In the region of 900-1000 bar, the improvement becomes marginal and BSFC starts to deteriorate at a given NOx level due to the need to retard the timing.

For more highly rated turbocharged engines of 1litre/cylinder, the minimum carbon fraction has been found to occur with injection pressures of 1100-1200 bar at rated power (Figure 4). The minimum achievable carbon fraction tends to be lower on the open chamber operated with higher injection pressures but full load carbon fraction acceptable for future off-highway standards can be obtained with injection pressures of around 1000 bar at the nozzle.

The position and inclination of the nozzle can have an important effect on the emissions. Figure 5 shows the effects of moving the nozzle from the standard position (4mm offset, 20° inclination) to a more central position (2mm offset, 10° inclination) on a turbocharged and aftercooled engine with a re-entrant chamber. Swirl level, combustion bowl and injection system were identical in both cases.

The results of a timing response comparison at rated power and peak torque show an improvement in smoke and particulates, especially at the peak torque condition, for the more central, less inclined nozzle. These effects are due to the improved nozzle tip and sac geometry which makes the distribution of fuel sprays more symmetrical about the chamber.

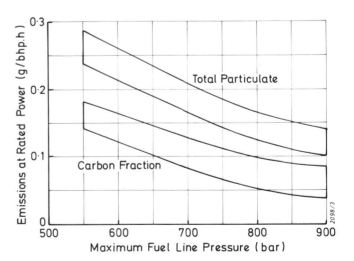

Figure 3   *Effect of injection pressure on particulates and particulate carbon at rated power for naturally aspirated DI engines.*
Air/fuel ratio 25-26:1
Timing set for 4.5-5 g/bhph NOx

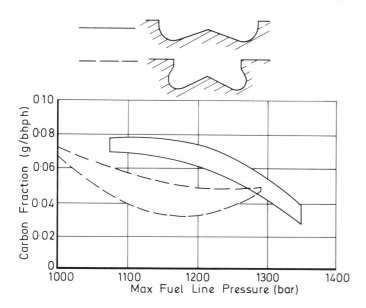

Figure 4   Effect of injection pressure on particulate carbon at rated power for a medium duty DI engine. (Reference 3)

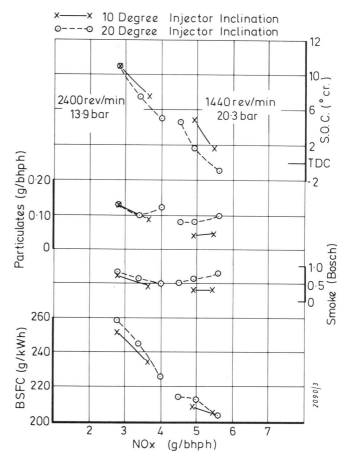

Figure 5   effect of injector location on full load performance (Reference 12)

4. EMISSION CONTROL STRATEGIES

The foregoing principles are applicable to all DI engines but the emphasis on each aspect of the technology depends on the engine type, application and the emission goals. Figure 6 illustrates CARB's proposed emission standards for 1996 and 2001 and shows the current development status of various prototype engines developed by Ricardo.

4.1 Naturally Aspirated DI Engines

This class of engine still constitutes a very large part of the off-highway engine sector. Typically these engines have a swept volume of around 1 litre/cylinder, a power output of about 15 kW/cylinder, and are produced in families of 3, 4 and 6 cylinders. Most are water-cooled, some are air-cooled. Fuel consumption, reliability, durability are important features but equally unit cost must be very competitive and thus relatively inexpensive unit injection pumps driven by the engine camshaft are widely used. Alternatives are in-line pumps or rotary pumps (for 4 and 6 cylinder engines).

CARB 1996 standards can be met by naturally aspirated (NA) DI engines, as shown in Figure 6, using the approach indicated in Table 3.

Combustion system design is likely to be subject to the limitations of the conventional 2-valve/inclined injector layout and moderate injection pressure capability. The re-entrant combustion system can be used in conjunction with injection pressures of 600 bar at the nozzle to give a good compromise over the speed range but an injection timing advance with speed is likely to be needed in most cases.

In this engine class the problem of part load carbon emission control is much greater, so that combustion chamber layout must be subject to minimum compromise, with centralised location of the bowl and injector, and minimised injector inclination.

At the CARB 1996 level, derating will not normally be required, depending on the engine's volumetric efficiency, hence ratings will be 7.5-8.5 bar BMEP at peak torque and 7-7.5 bar BMEP at rated speed.

The CARB 2001 levels present a significantly stiffer challenge for NA DI engines. Basic engine characteristics, such as volumetric efficiency and mechanical losses become more important. Tuned, and possibly variable length intake systems can be used to optimise volumetric efficiency at the two critical engine speeds in order to reduce full load smoke and/or maintain ratings. At the same time, mechanical and pumping losses need to be minimised in order to maximise useful work and thereby reduce brake specific emissions.

Injection pressures of around 1000 bar at the nozzle will be needed to reduce smoke at full load and part load carbon emissions. Most current designs of rotary pumps are unable to deliver this pressure but the development of high pressure rotary pumps appears promising. Their use is currently restricted to engines up to about 140 kW due to fuelling limitations. Alternative systems are in-line

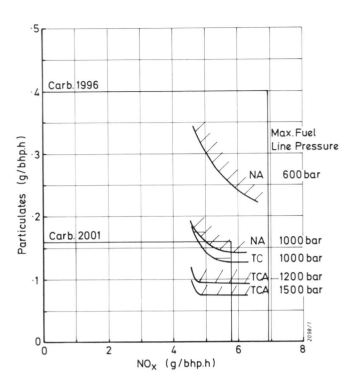

Figure 6 *Particulates v NOx trade-off over CARB 8 mode prototype DI engines Fuel .05% sulphur*

TABLE 3
STRATEGY FOR MEETING CARB OFF-HIGHWAY STANDARDS - TC DI ENGINE
X needs attention   XX critical feature

Feature	CARB 1996	CARB 2001
Rating	X	XX
Volumetric efficiency	X	XX
Friction	X	XX
Combustion chamber	re-entrant	re-entrant
Swirl ratio	intermediate	intermediate
Injection pressure	550-650 bar	900-1100 bar
Injector position	X	XX
Four valves/cyl	-	X
Timing control		
Speed advance	X	XX
Load control	-	X
Injection rate control	-	X
Injection rate shape	-	X
Nozzle type	minisac/VCO	minisac/VCO
Compression ratio	18-19:1	19-20:1
Oil consumption	0.15%	0.1%
Oxidation catalyst	-	X
Fuel	-	X

TC DI ENGINE
- Rating not so critical
- Wastegate needed for wide speed range
- Compression ratio 18-19:1 (Pmax limit)
- Aftercooler desirable for CARB 2001

pumps or unit pumps, both of which will require flexibility of timing control with respect to speed and load. In the ideal case, variable rate of injection would also be beneficial but most systems of this type would add too much cost.

Injector inclination and offset will need to be minimised, and serious consideration should be given to the use of a central injector and four valves, which provides the best possible combustion configuration. This adds cost but may be acceptable on engines produced in high volumes. It may also be a better solution overall to use a four valve, 4-cylinder, NA engine than the alternative of a turbocharged and possibly aftercooled 3-cylinder engine for the 50-70kW range.

Assuming the optimum match between combustion and injection systems, good control of lubricating oil in the particulate and low sulphur fuel (0.05% sulphur maximum), the NA DI engine is expected to be just inside the CARB 2001 standards. Depending on the soluble organic fraction of the particulate, an oxidation catalyst may enable the particulate limit to be met with an adequate margin [Reference 18].

At the CARB 2001 level, derating is likely to be required to avoid excessive smoke at the retarded timings required (Figure 7), hence ratings will be 7-7.5 bar BMEP at peak torque and 6-6.5 bar BMEP at the rated speed. The adoption of 4 valves/cylinder and central injector would enable any derating to be minimised.

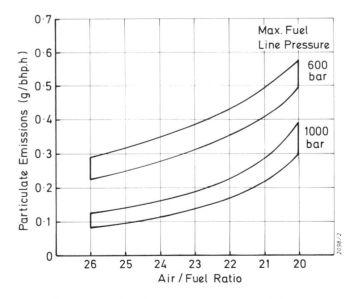

Figure 7 *Effect of air/fuel ratio on particulate emissions at rated power on naturally aspirated DI engines. Timing set for 4.5-5 g/bhph NOx*

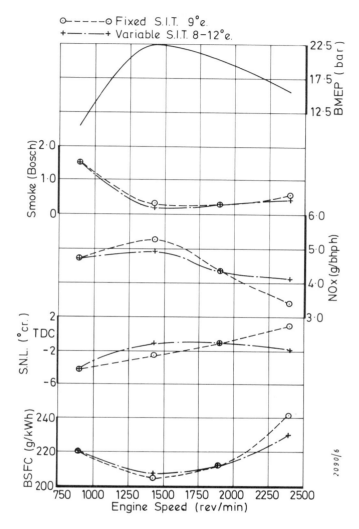

Figue 8   *Simulated speed advance with RP46 fixed prelift pump
(Reference 12)*

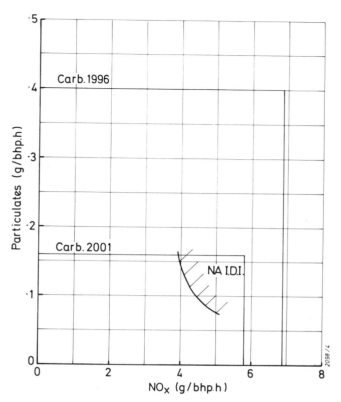

Figure 9   *Particulates v NOx trade-off over CARB 8 mode
prototype IDI engines fuel .05% sulphur*

shown in Figure 6. An example of the level of performance and
emissions which can be achieved on a 6 litre 6-cylinder engine with
an in-line pump and air-to-air aftercooling is shown in Figure 8. This
figure also indicates that fuel consumption can be improved by about
4% at the rated power by means of a timing advance.

In larger engines, alternative fuel injection systems may be
considered in some applications where the extra cost can be borne
and where, for example, electronics are already available on the
vehicle. The lowest particulate emission level in Figure 6 was
achieved with unit injectors which were capable of producing an
injection pressure of 1500 bar at the rated speed.

4.2 Turbocharged DI Engines

Many of the above arguments also apply to turbocharged versions
of the engines referred to in section 4.1, ie in the intermediate power
category, typically rated around 20 kW/cylinder.

CARB 1996 limits should be achievable but CARB 2001 limits
again present a challenge. The current approach to meeting CARB
2001 limits is shown in Table 3.

4.3 Turbocharged and Aftercooled DI Engines

High power versions of engines of 1 litre/cylinder with
aftercooling have been developed for off-highway applications and
these have demonstrated the ability to meet CARB 2001 levels, as

4.4 Indirect Injection (IDI) Diesel Engines

The development of industrial IDI diesel engines with Ricardo
Comet V combustion chamber has shown that prototype,
multi-cylinder NA engines with a swept volume of 0.4-0.9
litre/cylinder are capable of meeting the CARB 2001 limits (Figure
9) with little loss of performance or fuel economy compared with
the baseline IDI engine.

The key to the development lies in the precise matching of the
many parameters which affect combustion in the swirl chamber.
Once this has been achieved, a suitable timing plan is needed which
ideally will have both speed advance, load control and cold start
advance. The rotary injection pump is normally able to provide the
required timing plan without the need for electronics.

For 4 cylinder NA engines therefore, having a power of 30-60 kW, the IDI engine could represent a relatively low cost way of meeting CARB 2001 standards. Compared with the NA DI engine meeting Stage 2, the fuel consumption penalty would be about 10-15% at full load. At these ratings, thermal loading is not a limitation on water cooled engines.

5. CONCLUSIONS

Definitive exhaust emissions regulations for off-highway engines are not yet available but it is clear that moves to introduce regulations are underway in the US, Europe and Japan. It is likely that CARB's two stage approach will be adopted in California for engines over 175 bhp, starting in 1996, and spreading to all off-highway engines in the US depending on the outcome of EPA's deliberations. Japan appears to be moving towards the CARB approach also starting in 1996 or 1997. Europe is still at an early stage of discussion between the manufacturers represented by EUROMOT and the European Commission.

Apart from the debate over the limits, the "reference" test cycle is likely to be the CARB 8-mode, designated as cycle C1 according to ISO Draft Proposal 8178, but alternative cycles may be assigned to engines used in different applications.

In this paper, the principles of DI combustion system design have been reviewed, concentrating on the required injection system features and the related choice of the combustion system.

The potential of naturally aspirated (NA), turbocharged (TC) and turbocharged and aftercooled (TCA) DI engines to meet CARB 1996 and 2001 limits has been discussed. NA and TC DI engines have the potential to meet CARB 1996 limits using a relatively simple and low cost fuel injection system without loss of performance or fuel economy.

CARB 2001 limits present a challenge to the low cost NA and TC DI engines. Injection systems will need to be capable of about 1000 bar at the nozzle with flexible injection timing. On existing NA DI engines with 2 valves/cylinder and significant offset of the injector, it is unlikely that CARB 2001 limits can be achieved with an adequate margin unless major changes are incorporated such as the use of a central injector with 4 valves/cylinder and also an oxidation catalyst.

TCA DI engines have a better potential for meeting CARB 2001 limits without loss of performance and economy provided a suitable injection system is used.

For the lower power engines (30-60 kW), the NA IDI engine offers a relatively low cost route to achieve CARB 2001 limits.

ACKNOWLEDGEMENTS

The authors would like to acknowledge the support of their colleagues, notably Mr J R Needham, Mr A J Nicol and Mr S A Faulkner, and to thank the Directors of Ricardo for permission to publish this paper.

REFERENCES

1. Doyle D M, Needham J R, Faulkner S A, Freese R G
    "Application of an Advanced In-Line Injection System to a Heavy Duty Diesel Engine"
    SAE 891847

2. Needham J R, May M P, Doyle D M, Faulkner S A, Ishiwata H "Injection Timing and Rate Control - A Solution for Low Emissions"
    SAE 900854

3. Doyle D M, Faulkner S A, Needham J R, Broome D, Wotton C R N "Optimising the Six Litre Diesel for Low Emissions"
    SAE 905073

4. Doyle D M
    "Direct Injection Combustion Principles and Practice for Low Emissions"
    CIMAC 1991

5. Treiber P J H, Sauerteig J E
    "Present and Future European Exhaust Emission Regulations for Off-Road Diesel Engines"
    SAE 911808.

6. Timoney D J
    "A Simple Technique for Predicting Optimum Fuel-Air Mixing Conditions in a Direct Injection Diesel Engine with Swirl"
    SAE 851543

7. Timoney D J
    "Smoke and Fuel Consumption Measurements in a Direct Injection Diesel Engine with Variable Swirl"
    SAE 851542

8. Ball W F
    "Some Effects of Injection Pressure on Quiescent Direct Injection Diesel Combustion"
    Ricardo DP 80/453

9. Nishizawa K, Ishiwata H, Yamaguchi S
    "A New Concept of Diesel Fuel Injection - Timing and Injection Rate Control System"
    SAE 870434

10. Stumpp G, Polach W, Muller N, Warga J
    "Fuel Injection Equipment (FIE) for Heavy Duty Diesel Engines for US 1991/1994 Emission Legislation"
    SAE 890851

11. Frankl G, Barker B G, Timms C T
    "Electronic Unit Injectors"
    SAE 885013

12. Doyle D M, Faulkner S A, Nicol A J
    "The Role of the Fuel Injection System in Tomorrow's Low Emission DI Engine"
    Institute of Mechanical Engineers, 1992

13. Miyaki M, Fujisawa H, Masuda A, Yamamoto Y
    "Development of New Electronically Controlled Fuel Injection System ECD-U2 for Diesel Engines"
    SAE 910252

14. Shimoda M, Funai K, Shimokawa K, Otani T, Joko I
    "Application of a Heavy Duty Diesel Engine to Future Emission Standards"
    SAE 910482

15. Racine R, Miettaux M, Drutel Y, Heidt J
    "Application of a High Flexible Electronic Injection System to a Heavy Duty Diesel Engine"
    SAE 910184

16. Russell M F, Young C D, Nicol S W
    "Modulation of Injection Rate to Improve Direct Injection Diesel Engine Noise"
    SAE 900349

17. Bird G L, Duffy K A, Tolan L E
    "Development and Application of the Stanadyne New Slim Tip Pencil Injector"
    IMechE Seminar on Diesel Fuel Injection Systems, Solihull, 1989

18. Porter B C, Doyle D M, Faulkner S A, Lambert P, Needham J R,
    Andersson S E, Fredholm S, Frestad A
    "Engine and Catalyst Strategies for 1994"
    SAE 910604

ICE-Vol.18, New Developments in Off-Highway Engines
ASME 1992

THE DEVELOPMENT OF THERMAL SPRAY COATING TECHNOLOGY IN RESTORING LARGE BORE POWER CYLINDERS

David B. Aldag and Keith Kempton
Exline Incorporated
Salina, Kansas

Mitchell Dorfman
Metco Division
Perkin Elmer Corporation
Westbury, New York

Havery H. McAlexander
ANR Pipeline
Enterprise, Kansas

James B. Culp
Kaydon Ring and Seal Inc.
Baltimore, Maryland

ABSTRACT

For several decades, the power cylinders of large bore engines have been restored to like new dimension, with longer lasting wear surface by a porous chromium plating method. Efforts are now underway to develop a thermal spray coating process as a substitute for chromium plating.

This paper describes the compounds that were evaluated for the thermal spray coating processes. Results of wear tests against iron and against selected power ring materials are described. The process of moving from laboratory tests to the field tests is covered, as is a vision of what may come.

INTRODUCTION

It has been a practice for several years to apply a porous chromium plate to the internal bore of power cylinders operating in large bore engines. The process is regarded as a way to restore a worn bore surface without the expense of a new cylinder or liner. Also, chromium plating has been noted as providing a wear surface with long life characteristics in large bore engines. For example, a chromium plated power cylinder liner in an Ingersoll Rand KVG engine may wear 0.020" (0.50 mm.) on the diameter in 100,000 hours operation. The process has always had some shortcomings; the expense of applying coating thickness in excess of .020" (0.50mm), increasing costs of the repair, the lower cost of replacement liners, and lately, the cost of complying with environmental regulations. It seemed appropriate to look into other ways to restore the worn surface of large bore power cylinders. Coatings that have been applied with thermal sprays have been successfully operated in gas compressor cylinders, so the same process was considered for power cylinders.

This paper is concerned with the application of a thermally sprayed coating to restore the bore of power cylinders. The selection process is described and the first reports of field tests are presented. Four separate business entities are involved in this program; a repair shop that offers chromium plating now, a supplier of thermal spray equipment and materials, a ring manufacturer, and a pipeline operator of large bore engines.

The supplier recommended that a High Velocity Oxygen Fuel (HVOF) thermal spray process be utilized. The technology utilizes extremely high kinetic energy and controlled thermal energy output to produce very low porosity coatings that exhibit very high bond strength, fine as-sprayed surface finish and low residual stresses.

The process operates with an oxygen-fuel mixture, that is ejected at speeds approaching 4500 ft/sec (1350 m/sec) and ignited.

Combustion temperatures approach 5000 $^\circ$F (2760 C) and the flame forms a circular cross section. Powder is injected axially into the supersonic stream where acceleration and heating take place. The molten material is blasted on to the part to be treated.

The general test process was:
A. Select the materials that may be used.
B. Screen them by comparing wear tests on the selected materials with concurrent wear tests on the porous chromium plating.
C. Select appropriate ring materials.
D. Confirm the ring compatibility with the coating by further wear testing.
E. Confirm the process and metallurgy by applying the selected coatings with the equipment in the repair shop.
F. Select a four cycle, low BMEP engine with a continuous bore liner. (Ingersoll Rand KVG-12 gas fuel, spark ignition)
G. Thermally spray liners in the repair shop, using two materials, at two or more thicknesses.
H. Perform field tests with the liners, two coatings, two ring materials.
I. Extend the project, depending upon success, to different diameter liners, ported liners, thicker coatings, and sub-strate coatings.

MATERIAL SELECTION

The repair shop prepared a sample plug of typical chromium plating for the coatings supplier. Twelve materials and the chromium plated plug were run against a machineable cast iron ring to measure wear and friction. The Dow-Corning Model LFW-1 block and ring wear test machine was used for the tests. Friction factor, wear, hardness, weight change, and a basic description of each sprayed coating and the chromium plate sample is shown in Table 1. As a result of these tests the supplier recommended four materials for further study. These four materials were C, H, L, and P. These four materials were among those with the least scar width. Physical properties are described in Table 2. In comparison, microhardness for porous chromium plating is approximately DPH300 750-850.

The four coatings were then tested for wear against the two ring materials (to be described later in this paper) that were suggested for the top ring in the ring pack. Scar width, weight loss and coefficient of friction were some of the criteria used to select the best materials.

TABLE 2   COATING PROPERTIES -- SELECTED MATERIALS

Material	C	H	L	P
Macrohardness, Rc	40-45	45-50	45-50	25-30
Bond Strength, psi	9,000	8,500	8,500	8,000
Thickness Limit, inch	0.060	0.050	0.050	>0.150
Metalurgical Characteristics				
% Oxides	1	20-22	<1	17-18
% porosity	1	1	1-2	<1
% unmelted particles	5-7	2-3	8-10	10
Surface Texture				
as sprayed, microinches aa	250-350	250-350	250-350	275-375
as honed, microinches aa	9-12		9	
Microhardness, DPH 300	550-650	625-725	650-750	200-300

Table 3 illustrates the test results of these two sets of ring materials. Another test involving chromium plating (Rc 59) was run for three of the coating materials. Those results are also shown in table 3. It can be seen that weight change and the scar width for materials C and L are lower than the other two materials when run against the iron ring materials. Therefore, materials C and L were chosen to be applied for the field test.

RING SELECTION

The ring supplier selected two ring packs that were expected to run against these four materials. Low porosity of the coatings and the potential for extending this project to higher loaded engines was considered.

The oil rings of both packs carried less unit loading (120 psi) than normal to enhance the oil distribution in the low porosity cylinder. Both oil rings were described as conformable grooved oilcutter rings.

The next two compression rings were identical, described as taper faced, butt cut rings made of low alloy grey cast iron.

TABLE 1   WEAR TEST RESULTS AGAINST GREY CAST IRON (Rb=98)

Material	Friction Break-in	Friction Steady State	Scar Width (mils)	Ring Weight change (mg.)	Coating Weight change (mg.)	Coating Hardness Rc	Coating Components
A	0.12	0.12	33-35	-1.2	-0.5	59	Chromium plating, baseline
B	0.13	0.15	48-50	-2.7	-0.8	41	Iron, Chromium, Carbon
C	0.14	0.14	45-50	-1.6	+0.4	42	Iron, Chromium, Molybdenum, hardfacing alloy
D	0.16	0.16	52-54	-2.1	-0.4	51	Cast Iron
E	0.14	0.15	68-71	-5.7	-1.2	49	Cast Iron, Iron Oxide
F	0.13	0.17	59-61	-2.0	-0.6	50	Cast Iron/Iron Oxide Cermet
G	0.12	0.14	62-66	-2.2	+0.2	50	Pre-oxidized Cast Iron
H	0.12	0.13	46-48	-1.0	+0.2	50	Cast Iron/Iron Molybdenum
J	0.11	0.13	45-48	-1.3	-0.6	52	Iron, Molybdenum
K	0.11	0.13	40-43	-2.3	-0.1	53	Iron, Molybdenum, oxidized
L	0.14	0.14	32-37	-0.6	+0.1	48	Iron, Molybdenum, self fluxing Nickel base alloy
M	0.16	0.16	70-75	-1.1	+0.5	47	Pre-alloyed Cast Iron Molybdenum powder
N	0.13	0.16	53-58	-2.3	-0.8	53	Pre-oxidized Cast Iron, Iron, Molybdenum
P	0.11	0.12	39-42	-0.8	-0.2	30	Iron, Chromium

10mils=0.010"=0.254mm   1 mg.=0.000035 ounces approx.

114

The only difference in each set was in the top ring. Both top rings are described as angle cut, crowned, filled groove. The fill material in both rings is a blend of iron oxide and silicate of soda, used to carry lubricant and prevent scuffing. The difference in the two top rings is the base material. One ring of made as grey iron with some alloys added for strength, and the hardness is approximately Rc 34. The top ring on the other set is made as a nodular iron (Rc 46) and is almost as hard as the liner materials. The nodular iron is stronger and harder than the alloyed grey cast iron. The nodular iron ring has been proven in operation in chromium plated cylinders.

SHOP APPLICATION

Applying thermal coatings on the inside of a bore presents some unique problems. The bore, or liner, must be swung as if it is in a lathe to present all of the bore surface to a stationary nozzle. The residue materials and gases can flow out one or both ends, become entrained in the flame, or leave deposits on parts of the cylinder. Therefore, coupons were installed inside a test liner and Materials C and L were sprayed at the repair shop. These coupons were examined to confirm the procedures, the metallurgy and physical characteristics. Microstructure examination confirmed that the procedures were acceptable and that the project could proceed.

Also, at this time, it was discovered that bond strength between the sprayed alloys and well worn liners that had been restored with iron plate was unsatisfactory. It was decided to postpone the development of improved spray techniques for iron plated cylinders until there was a reasonable chance for success in the balance of the project.

FIELD TEST

With the materials selected, the rings selected, and the spray process confirmed, it was time to move ahead and spray some liners for actual field trials. Five liners were sprayed at the repair shop in August 1991, using the two materials, in three thicknesses. The length of the bore was 41" (1.05 m) and the diameter was 15.25" (387 cm.). The liners were mounted in a fixture which allowed relatively free flow of gases from both ends of the liner. The HVOF gun was mounted with two auxiliary streams of air to purge the overspray from within the liner bore. The thermal coating was applied by utilizing traverse in both directions. It was reported that this arrangment exhibited less overspray than most thermal coatings normally applied in the shop.

TABLE 3 WEAR TEST RESULTS, SELECTED MATERIALS

Material	Wear Ring Material Rc *	Friction Break-in	Friction Steady State	Scar Width (mils)	Ring Weight change (mg.)	Coating Weight change (mg.)	Coating Hardness Rc	Coating Components
A	34	0.16	0.15	64—67	−0.5	−0.7	59	Chromium plating, baseline
A	46	0.15	0.15	67—70	−1.2	−1.4	59	
C	34	0.12	0.12	45—48	−0.1	−0.3	42	Iron, Chromium, Molybdenum, hardfacing alloy
C	46	0.13	0.13	36—39	−0.8	−0.1	42	
C	59	0.17	0.16	76—79	−10.2	+1.6	42	
H	34	0.14	0.15	56—59	−0.2	−0.7	50	Cast Iron/Iron, Molybdenum
H	46	0.14	0.16	61—63	−0.5	−0.4	50	
H	59	0.13	0.12	26—30	−0.6	0.0	50	
L	34	0.12	0.12	35—37	+0.2	−0.3	48	Iron, Molybdenum, self fluxing Nickel base alloy
L	46	0.12	0.12	31—34	−0.5	−0.1	48	
L	59	0.17	0.14	59—62	−1.9	−0.4	48	
P	34	0.13	0.14	67—69	−0.6	+0.2	30	Iron, Chromium
P	46	0.14	0.14	68—70	−0.6	−0.5	30	

10mils=0.010"=0.254mm   1 mg.=0.000035 ounces approx.

* WEAR TEST RESULTS AGAINST GREY CAST IRON (Rc=34)
  WEAR TEST RESULTS AGAINST NODULAR IRON (Rc=46)
  WEAR TEST RESULTS AGAINST HARD CHROMIUM PLATE (Rc=59)

The hone operator's observations of excess heat, hone chatter, stone clogging and sound led to a refined honing process. The types of stones, the honing speeds, and the honing pressures were recorded for future production. There is still opportunity for improvement in control of the amount of material deposited so that honing can be minimized.

The liners were installed at the pipeline operator's station in a KVG engine. This is a gas fueled, spark ignition engine driving integral compressors. Horsepower is 1320 (985 KW.) at 330 rpm. The operator recommended an inspection scheme to detect damage before it became catastrophic. Frequent inspections using a borescope to inspect the cylinder walls were scheduled, with a reduction in frequency after successful running time. Crankcase oil analysis frequency was increased and a visual inspection of the cylinder walls was scheduled after a successful one month run. The plant operators increased their frequency of vibration analysis of the engine.

It should be noted that a different oil formulation was used for crankcase make-up when the engine was started. The formulation was changed in response to the needs of another engine in the plant, not because of the test engine. The old oil was an ashless detergent oil with oxide inhibitors and anti-wear compounds. The new oil was a low ash oil with more detergent properties and more oxidation and nitration inhibitors. Oil is not taken out of the crankcase to lubricate the compressor cylinders. This change is not expected to have any bearing on this project.

The arrangement of materials, rings, coating thickness and liner location is described in Table 4. The other cylinders of the engine were left undisturbed.

TABLE 4   POSITION OF MATERIALS AND RINGS IN KVG ENGINE

Material	Cylinder	Coating	Finish RA	Liner Hardness	Ring Hardness	Coating Components
C	#3	0.056"	9—10	Rc42	Rc34	Iron, Chromium, Molybdenum, hardfacing alloy
C	#7	0.020"	10—12	Rc43	Rc46	Iron, Chromium, Molybdenum, hardfacing alloy
C	#11	0.040"	10	Rc42	Rc34	Iron, Chromium, Molybdenum, hardfacing alloy
L	#5	0.040"	9	Rc50	Rc34	Iron, Molybdenum, self fluxing Nickel base alloy
L	#10	0.020"	9	Rc48	Rc46	Iron, Molybdenum, self fluxing Nickel base alloy

10mils=0.010"=0.254mm

Within a few hours of start-up the cylinders were all examined with a borescope. An oil analysis at 180 hours indicated an increase in iron and chromium. The levels have remained at about the same point in the subsequent operating hours. The engine is automatically controlled to operate at 100-105% torque and varying speed. Speeds may have been as low as 250 rpm at times during this test.

At about 235 operating hours, a vibration analysis of the cylinders indicated that an unusual noise was developing during the firing of the #7 cylinder (material C). This pattern became intermittent at about 700 operating hours. From start-up, the borescope inspections showed that some scuffing was developing in the #7 cylinder. One of the first areas, and ultimately the largest, started at the top of the oil ring travel. It extended as a vertical streak in the direction of the oil ring travel. This damage and other streaks grew progressively wider. In the undamaged parts of the liner, the hone marks were still visible and no disturbance or wear in the ring reversal area was noted. This liner was removed from service when the "heads-off" inspection was in progress. Visual inspection indicated that the bond between the parent metal and the sprayed metal was still intact. A more complete examination is scheduled.

Borescope examination of cylinders #3 and #11 (Material C) indicated some distress by the appearance of vertical streaks in the ring travel area. Some of these streaks grew larger during operation. In the other parts of the cylinder the wear seemed to be minimal, as the hone marks were still visible. During "heads-off" inspection (900 hours operation) these streaks seemed to be smoothed over. Polishing with emery cloth would have cleaned them up.

Borescope inspection of Cylinders #5 and #10 (material L) showed that the hone marks had disappeared and a ring reversal pattern had developed in the top of the liners. At the "heads-off" inspection, these liners measured 0.002" wear in the ring reversal zone. The engine vibration patterns did not indicate anything to be concerned about. While there were some vertical scratches apparent in the #10 cylinder, there were no signs of on-going scuffing in either cylinder.

Cylinder lubrication was adequate. It was decided to continue operation of the engine and make another inspection at 6500 hours operation, (August 1992).

OBSERVATIONS
Wear in the cylinders with material L is more than what is expected and the material may not be a good substitute repair. The "heads-off" inspection is not considered to be conclusive in this regard. Observations at the 6500 hour period will be more conclusive about wear.

Material C appears to be an exceptional material for wearing quality. However, it is still a questionable candidate because of the tendency to score. Further examination of the liner that has been removed from the engine may offer some clues to resolving the scoring issue.

A third material, H, may have some possibilities. It was tested along with materials C and L against a chromium surface wear ring. The measured scar widths and weight loss were less than any other combination in the lab wear tests. For this material to be utilized, the ring pack would have a top ring with a hard chromium surface. These possibilities are being explored.

WHAT'S IN THE FUTURE
There are a lot of avenues for further development. This project can be expanded to include ported two-cycle cylinders. Also larger and smaller bores should be attainable. Thicker coatings with less expensive substrates are possible. This would allow the restoration of cylinders that are currently worn too much to apply chromium plate. It is expected that application rates in the thermal spray process can be increased significantly. It would seem reasonable to suggest that coatings of this nature would also be applicable to new cylinders if improved wear could be attained. And finally, there is some thought about restoring cylinders at the engine site without bringing them into a shop.

ACKNOWLEDGMENTS
Participants in the project other than the Authors are Jerry Exline, Exline Inc; Francis Montgomery, Metco Division of Perkin Elmer; Tom Gimse and Harold Baker of ANR Pipeline; and the Kaydon Ring and Seal Company. Their contributions are gratefully acknowledged.

ICE-Vol.18, New Developments in Off-Highway Engines
ASME 1992

FIELD TESTING TO VALIDATE MODELS USED IN
EXPLAINING A PISTON PROBLEM IN A LARGE DIESEL ENGINE

M. J. Graddage
Ricardo Consulting Engineers Ltd.
Shoreham-by-Sea
West Sussex, United Kingdom

F. J. Czysz
Pennsylvania Power & Light Co.
Allentown, Pennsylvania

A. Killinger
MPR Associates Incorporated
Washington, D.C.

ABSTRACT

Two crankcase explosions occurred within one month in diesel engines which drive large emergency generator sets at a nuclear power plant in Eastern Pennsylvania. As a result, the electric utility conducted an extensive investigation to determine the root cause(s) of the problem. Initial inspections confirmed that the crankcase explosions were the result of pistons and liners becoming overheated. The technical challenge was to establish why the pistons and liners were overheating when other engines of the same type did not appear to have the problem in the same duty.

Analytical models of piston motion, engine start and run thermodynamics, and a finite element analysis of piston distortion during engine start and load transients were developed. Preliminary work with these models predicted a feature of the piston design which could adversely affect lubrication conditions during a rapid start and load transient. Final input data to refine the models was needed and this was obtained from tests carried out on a similar diesel generator operated by a municipality in Iowa.

This paper describes the successful accomplishment of the field tests using state-of-the-art instrumentation and recording equipment. It also shows how the modelling and test work identified wear at certain locations on the piston skirt as the origin of distress leading to the crankcase explosions. Unfavourable engine starting and loading conditions as well as less than desirable piston skirt-to-liner lubrication conditions in the engines at the nuclear power plant have been identified as the root causes and corrective action has been initiated.

1. INTRODUCTION

In a one month period, crankcase explosions occurred in two of the five emergency diesel generator set engines at the Susquehanna nuclear power plant in Eastern Pennsylvania operated by the Pennsylvania Power and Light Company (PP&L). The engines were being subjected to a routine monthly test at the time to demonstrate their continued availability, a procedure which involved starting the engines from stand-by conditions and bringing them up to full load within 90 seconds. The engines, 16 cylinder units rated at 4MW, are a type licensed by the U.S. Nuclear Regulatory Commission for this duty and are installed at eight other nuclear power plants. At these nuclear power plants, as well as other stationary and marine applications, the engines have a generally good reliability record.

A preliminary investigation (1)* quickly established that the crankcase explosions were a result of pistons and liners becoming overheated due to localised scuffing on the upper diameter of the piston skirt and liner on the anti-thrust side (the side opposite that carrying the major side forces due to conrod angularity on the power stroke). However, no defect in the pistons, liners or other causal factors could be easily identified which would clearly explain why the scuffing should have occurred in this unusual position. Up until this time crankcase explosions had not been experienced by any of the other engines operating in the same duty. Whilst introducing changes to operating procedures and interim modifications designed to improve the lubrication conditions of the piston and liner, PP&L therefore also initiated an in-depth investigation to establish the root cause of the problem.

This paper provides an outline of the work done to assist the electric utility to examine the operating conditions of the piston and liner inside the diesel engine. The paper describes how analytical modelling and field test work were combined to successfully identify the root cause of the crankcase explosions.

2. METHOD OF APPROACH

Because of the limited availability of the engines for test work at the nuclear power plant, the primary method of investigating the piston problem was analytical. A Finite Element (FE) model of the

* Numbers in brackets designate the references at the back of the paper

piston and liner was used to predict piston distortions and piston/liner contact pressures. The rapid start and loading sequence regularly employed at the nuclear power plant was considered a potential causal factor. Therefore, to show how the piston/liner contact changes through the operating cycle analyses were carried out for stand-by conditions, full load steady state and two intermediate points in the loading transient.

The starting point for the distortion and contact patch analysis was component temperatures and piston forces and attitudes. These were determined by the simulation of piston motion and component thermal conditions, both at steady state and throughout the start-up and loading transients. As this type of analysis had previously been carried out (with considerable success) only for steady load, steady temperature conditions, a programme of test work involving temperature and motion measurements was undertaken to provide validation for the results of the simulation.

The validation test work was carried out on a diesel generator operated by a municipality in Sumner, Iowa. Although this engine was not identical to the nuclear power plant engines it provided data for similar operating conditions. The actual conditions were then assessed using the analytical models with only a small amount of extrapolation.

3. PRELIMINARY MEASUREMENTS

Before undertaking the simulation work, a series of measurements were taken from the nuclear emergency diesel generator engines during typical start and running conditions. A comprehensive package of instrumentation and data

logging equipment was employed to measure and record in-cylinder pressure, injector needle lift and other more general temperatures and pressures, etc. The results of these tests provided baseline data for the analysis work but also highlighted a characteristic of the air starting system of the engine which caused compression and maximum cylinder pressures to be increased significantly during the first few cycles of a start, see Figure 1. This finding was considered significant as the loadings on the anti-thrust side of the piston produced by the cylinder pressure would be increased to near those seen at full load but, at a lower speed where lubrication and the beneficial effects of inertia forces (see Section 5.1) would not be so great.

Some imbalance in tuning of the nuclear power plant engine was also revealed by the measurements but neither this or any other unusual characteristics were identified which could be shown to be the sole root cause of the piston problem.

Once the baseline data had been recorded from the engine at the nuclear power plant further measurements were taken from an engine operated by a municipality in Iowa. This engine had been identified as potentially suitable for use as a validation tool as it was similar to those at the nuclear facility and employed the same pistons and liners. It was a 12 cylinder dual-fuel engine operating at 514 rev/min whereas the nuclear power plant engines were 16 and 20 cylinder models running on diesel fuel at 600 rev/min. The inertia of the generator was also higher than that of the nuclear power plant engines and the air starting system did not exhibit the same pressure 'boosting' characteristics. However, the full load bmep of the engines were similar and the tests, coupled with supporting piston motion analysis work, established that the engine could be modified

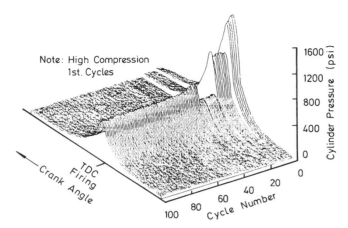

Figure 1   *Measured cylinder pressure of nuclear plant engines during start.*

and operated in such a way as to provide the data necessary to validate the computer models.

4. ENGINE TESTS TO VALIDATE THE MODELLING WORK

4.1 Engine

Some details of the main features of the engines at the nuclear power plant are given in Table 1 compared with those of the test engine.

4.2 Instrumentation

A comprehensive package of instrumentation was installed on the test engine to monitor start air system operation, engine performance parameters, component temperatures and piston motion. Details of some of the instrumentation and data logging systems employed are given in the following sections.

4.2.1 Component Temperatures Piston and liner temperatures were measured by thermocouples. Twenty were installed in a piston and seven in a liner, both components being installed in the same position in the engine.

Signals from the piston thermocouples were communicated to the data logging system through intermittent contact gear designed and manufactured by Ricardo. The thermocouple leads were fixed to the inside of the piston and brought down to contact pins on the lower edge of the piston skirt. At the bottom dead centre position these pins made contact with leaf springs attached to the bottom of the liner, see Figure 2. During the brief period of contact the signal from each thermocouple charged up a capacitor. The time constant of the capacitor circuit was chosen such that a steady voltage was maintained between contacts and it was this voltage that was logged. The liner thermocouples were wired directly to the data logger.

4.2.2 Piston Motion   Piston motion was investigated using eight Micro-epsilon eddy current proximity probes mounted in the

TABLE 1
MAIN FEATURES OF NUCLEAR POWER PLANT AND TEST* ENGINES

Bore:	13.5 ins
Stroke:	16.5 ins
Rated Speed:	600 rev/min (514 rev/min)
Cylinders:	16 at 45° included vee angle (12 cylinders)
Continuous rating:	4000kW electrical (2748 kW electrical)
Rated BMEP:	13.6 bar (14.6 bar)
Piston Type:	One piece cast iron, tin plated skirt
Connecting Rod Configuration:	Master and Slave

* Figures in parenthesis refer to test engine.

Figure 2   *Contact gear leaf springs on carrier ring*

liner. The probes were tuned to give an output proportional to the gap when only the skirt of the tin plated piston was opposite the probe. The passage of the rings and ring lands was not monitored. The main objective of the measurements was to monitor piston motion in the thrust/anti-thrust plane and, therefore, six of the eight sensors were in this plane (four on one side two on the other). These probes were distributed such that a minimum of three probes were adjacent to the piston skirt at any instant. This ensured that both angular and translational data were available at all times. Two probes were placed in the crankshaft axis direction to give an indication of motion in the orthogonal plane. Signals from the probes were input to a signal conditioning unit which linearised the probe output and corrected for probe temperature.

4.2.3 Data Logging Systems   Cylinder pressure, needle lift and proximity probe signals from the test cylinder were logged at each degree of engine crank angle with a Ricardo Configurable High Speed data Acquisition System for Engines (CHASE) (2). The CHASE system comprised a front end multi-channel acquisition unit, communicating with a PC which provided data processing, display, archiving and user interface functions. Ten channels of analogue data were acquired synchronously on a trigger from a

digital encoder coupled to the after end of the generator shaft.

For general purpose data acquisition of temperature, pressure etc. a system comprising of a number of Isolated Measurement Pods (IMPs) located adjacent to the engine, communicating with a PC via a serial data link was employed. Each IMP was a self contained multi-channel data acquisition unit, providing multiplexing, analogue to digital conversion (A/D), scaling, cold junction compensation and communication functions, all housed inside a robust enclosure.

Logging data every degree for every cycle of a test run was not necessary and would have produced impractically large quantities of data. In order to avoid this and to have data throughout each test run, a selective sampling technique was employed: The CHASE high speed logging system was configured to count engine cycles and was programmed with a map of which cycle numbers to record. A typical logging regime was the first 50 engine cycles followed by 5 cycles every 50 throughout the rest of the test. The general purpose logger was programmed to record data every second for the first part of test and then every 3, 10 or 30 seconds thereafter as appropriate.

4.3 Test Procedure
The engine was set up prior to test to run in the diesel only mode. The original dual fuel pumps were retained but the camshafts were retarded in order to produce in cylinder conditions which were closer to those of the nuclear power plant engines. Fuel injector nozzles were installed that were identical to those used in diesel engines at nuclear power plants.

The engine was run-in and then a series of tests were undertaken covering different motoring and starting conditions and zero to full power loading times varying from one hour down to only sixty seconds. A summary of some of the tests carried out is given in Table 2 and the variation in pre-start conditions achieved is summarised in Table 3.

For most of the tests a maximum power setting of 2900kWe was employed (5.5% overload on the normal full power of the test engine in the diesel mode). The intention was to ensure that in-cylinder conditions of the test engine were at least equal to or slightly more severe than those of the nuclear power plant engines.

4.4 Test Results
Typical plots of piston temperature versus time are shown in

TABLE 2
RANGE OF TESTS COMPLETED

Hours Since Last Run	Pre-Start Rack/Governor Setting	Idle Time	Rate of Loading	Remarks
18	0 mm 0.55	5 mins	0-2670 kW in 10 mins	'Normal' start
5 ½	45mm 0.65	40 secs	0-2600 kW in 60 secs	Rapid start from 'warm'
½	45 mm 0.65	1½ mins	0-2900 kW in 60 secs	Rapid start from 'hot'
16	45 mm 0.65	1 ½ mins	0-2900 kW in 60 secs	Rapid start from 'cold'
½	0 mm 0.55	5 mins	750 kW every15 mins	Slow start and load

TABLE 3
VARIATION IN PRE-START TEMPERATURES

		Max	Min
Piston (crown edge)	°C	100	25
Piston (crown centre)	°C	65	25
Liner (mid-stroke)	°C	60	25
Oil	°C	65	25
Coolant	°C	50	25

Figure 3. The three piston thermocouples shown cover the full range of readings from all twenty thermocouples. The crown temperatures react quickly to the application of load and reach maximum temperature within two minutes of full load being applied. The response of the skirt temperatures, represented by the bottom temperature in Figure 3, is much slower and is similar to that of the oil and coolant temperatures.

The temperature of the piston at five positions (from crown centre to skirt bottom) measured before start, during transient and at full power for two tests with widely different loading rates are compared in Figure 4. This plot gives an indication of the effect of loading rate on the temperature distribution within the piston. The upper plot shows that the pre-start temperatures of the two tests were a little different but confirms that the piston was at uniform temperature before each test. The lower plot shows that there was a similar temperature gradient throughout the piston at the end of each test. The centre plot shows a point where the crown centre thermocouple had reached a temperature approximately half way between the stand-by and steady state conditions during each test. The time after start at which this temperature was reached was different for each test due to the difference in loading rates. However, it can be seen from the plot of the other thermocouple readings at this 'mid point' that the temperature distribution throughout the piston differed for each test but by only a small amount (approximately 20°C). This finding shows that the thermal gradient down the piston is controlled primarily by crown temperature and not loading rate.

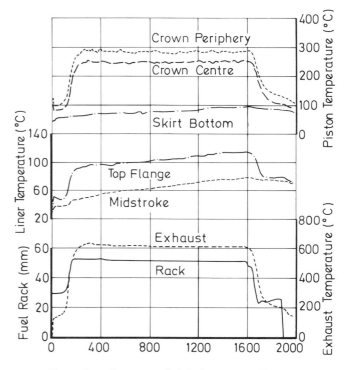

Figure 3   Data recorded during start and load test

Two of the seven liner thermocouple temperatures are shown in the middle plot of Figure 3. Although the top flange temperature changed quickly, the temperatures of the liner in that part of the cylinder swept by the piston skirt responded to the increase in load much more slowly and in a similar way to those of the upper and lower piston skirt. This suggests that liner/piston skirt clearances do not change greatly as the engine warms up (coefficients of thermal expansion are similar).

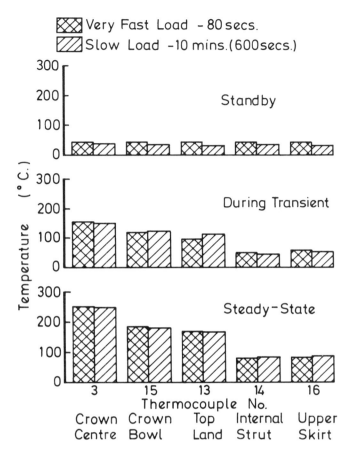

Figure 4   Comparison of piston temperature distribution for fast and slow loadings.

5. ANALYSIS WORK

In the following sections, selected results of the piston motion, thermal transient and distortion and contact patch modelling work carried out are shown and compared to the results of the test work covered in Section 4. The modelling methods employed are described in more detail in Reference 3.

5.1 Piston Motion

Piston secondary motion, tilt and side forces for use in the finite element study of piston/liner contact during engine start-up were calculated using a piston motion program (PISMOT). Input data for the program included component dimensions, masses, coefficients of friction and cylinder pressure. Calculations were carried out for a range of operating conditions and speeds and with a variety of cylinder pressure loadings. Piston motion and side forces were calculated for the whole of the 720 degree operating cycle at these conditions.

The results of the calculations confirmed that the difference in piston motion between the test engine at its rated speed of 514 rev/min and that of the nuclear power plant engines at their 600 rev/min rated speed was almost indiscernible and that on loadings quite small, see Table 4.

The predictions of typical full load and light load cycles of piston motion are compared in Figure 5. It can be seen that the predictions are in generally good agreement with the measured motions, particularly when gas pressure and hence piston side loading, is high.

It was noted that both prediction and measurement showed that the piston is held flat against the anti-thrust side for the whole of the compression stroke at full load due to the increased compression pressures produced by turbocharger boost at this load. At light load, the compression pressures are not high enough to keep the piston pressed hard against the anti-thrust side for the whole of the compression stroke, the piston being 'lifted off' the cylinder wall in the final half of the stroke by inertia forces. This effect is also well illustrated by the 3-D plot of measured piston motion shown in Figure 6. Here piston secondary motion is plotted every 100th cycle over a period in which the engine load was increased from 1500 to 2250 kW. It can be seen that piston motion changes during this load step due to the increased boost and resulting compression pressures. This plot also shows that piston secondary motion is very consistent from cycle to cycle at the same load.

Measured piston motion for the first 50 consecutive cycles recorded during a typical start of the test engine is shown in Figure 7. For the first 18 or so cycles, as the engine accelerates from rest and inertia forces are relatively low, compression pressures, even without the aid of boost from the turbocharger, are sufficient to hold the piston against the anti-thrust side of the cylinder during most of the compression stroke. However, as the speed of the engine rises the effects of inertia become more pronounced, the compression pressures can no longer hold the piston against the anti-thrust side for the whole of the compression stroke.

The effect of component temperatures on piston motion was negligible. The skirt warms up relatively slowly, at a similar rate to the liner and, therefore, the skirt to liner clearance does not change much as the engine warms up. Although the shape of the piston changes due to the temperature increase this does not appear to affect piston motion significantly.

The results of the piston motion measurements successfully validated the motion modelling work and hence validated the use of the forces calculated by the simulation in the examination of piston distortion and piston/liner contact patch.

TABLE 4
COMPARISON OF PISTON SIDE THRUST AT TWO SPEEDS

	Maximum Piston Side Thrust (kN)	
	600 rev/min	514 rev/min
Top of Skirt		
Major thrust side	34.2 @ 30	36.2 @ 30
Minor thrust side	13.0 @ 600	13.3 @ 700
Bottom of Skirt		
Major thrust side	22.0 @ 30	23.3 @ 30
Minor thrust side	5.2 @ 400	5.0 @ 700

Additionally, the difference in timing of the piston cross-overs is less than one crank degree.

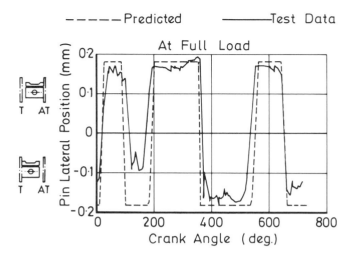

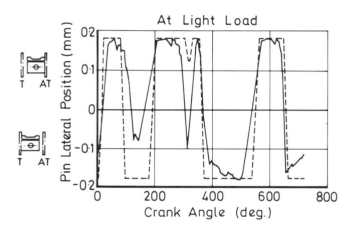

Figure 5   *Comparison of measured and predicted piston motion.*

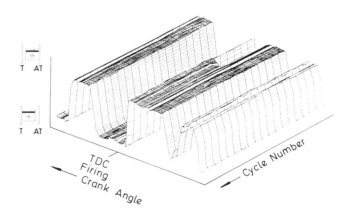

Figure 6   *Piston lateral movement every 100th cycle during a slow transient from 1500kW to 2250kW*

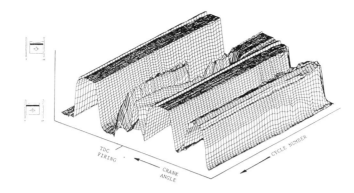

Figure 7   *Piston lateral movement for the 1st, 50 cycles of a typical fast start*

5.2 Transient Thermal Analysis

Before the distortion and contact patch analysis could be undertaken it was necessary to identify the 'worst case' time steps in the start up and loading transient and provide boundary conditions for these steps in addition to the steady state information more usually employed for the distortion analysis. Additionally, it was necessary to use modelling to compare the rate of thermal loadings of the test engine with those of the nuclear power plant engines to enable the validated analysis for the test engine conditions to applied to the nuclear power plant. Ricardo carried out the performance/thermal simulations using their IRIS program. (4)

The preliminary phase of the work indicated that temperature gradients and, therefore, distortions were greatest at full power, steady state conditions. Component temperatures rose from stand-by conditions to the full power state gradually without unusual excursions. This was confirmed by the test work, Figure 3. Therefore, three thermal condition time steps were identified for the preliminary distortion analyses; stand-by, full load and a point after start which was the earliest time that full fuelling had been applied to the test engine.

Once the engine test work was completed, validation of the preliminary analysis work and adjustment of the model to match the measured conditions was undertaken and a further intermediate time step was also identified for the distortion analysis. The predicted heat fluxes were modified to match the calculated temperature map with the piston temperature measurements and the resulting comparison between predicted and measured values is shown in Figure 8.

A comparison of the thermal response of test and the nuclear power plant engines using the models is shown in Figure 9. It can be seen that the objective of producing a slightly more severe thermal transient on the test engine than that seen at the nuclear power plant was achieved. The maximum steady state piston crown temperatures for the test engine condition were 20-25°C higher than normal operation at the nuclear power plant but the temperature distribution through the piston was very similar.

Validation of the transient thermal analysis confirmed the heat flux values used in the finite element model to determine distortions.

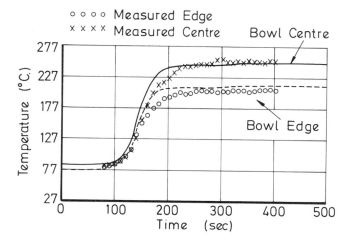

Figure 8   Comparison of predicted (lines) and measured (symbols) temperature histories for the test engine piston during transient.

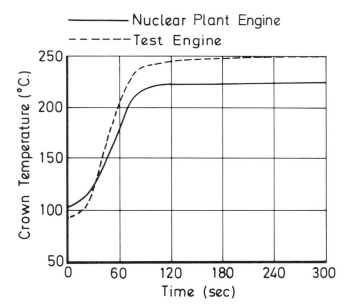

Figure 9   Comparison of predicted piston transient temperatures for nuclear plant and test engines

5.3 Distortion and Contact Patch Analysis

The objective of the distortion and contact patch analysis was to calculate piston/liner contact conditions and thereby assess the possibility of scuffing occurring during a start up and rapid loading transient.

The method employed was to use a 3-D FE model of the piston to assess the contact conditions with the piston flat against the bore near TDC, at the point of maximum side thrust as identified by the

motion analysis work. The following cases were examined in detail:

  A. Start up at 100 rev/min.
  B. Two transient conditions at 600 rev/min, full load. (Nominally 51 and 60 seconds after start)
  C. Steady state condition at 600 rev/min, full load.

The calculations show that the sidewall surface of the unloaded piston skirt is concave in the cylinder axis direction at stand-by temperature but rapidly changes to a better convex shape when the piston has reached final full load temperatures. Convex profiles assist the formation of hydrodynamic oil films in either direction of piston travel. The concave shape at stand-by conditions is a result of stresses caused by the pressing in of the wrist pin bush and differential expansion between the bush and piston materials.

The unloaded piston skirt profiles for the four conditions are compared in Figure 10. The corresponding contact patches under worst case anti-thrust side loading are shown in Figure 11. It can be seen how the concave profile at stand-by temperatures (a) will inhibit the formation of oil films on the compression stroke, because of the 'sharp' edge on the leading edge of the skirt. Conditions on the down stroke are much more conducive to oil film formation because the gradient at the leading edge is less. In addition, the bottom of the skirt is more flexible than the top, it will therefore more readily conform to the liner bore.

The unloaded skirt profile at the first intermediate time step (51 seconds) has a small lead-in at the top of the skirt. However, predictions for the piston under side loading, Figure 11 (b), show that the skirt has been flattened and that oil film generation would be poor. However, by the second intermediate time step (60 seconds), the skirt profile has improved and it becomes much better as the piston warms up to the full load steady state condition where the profile and contact patch are much more acceptable.

Calculations of peak skirt/liner contact pressure for the four cases were also carried out and these confirmed that contact pressures on

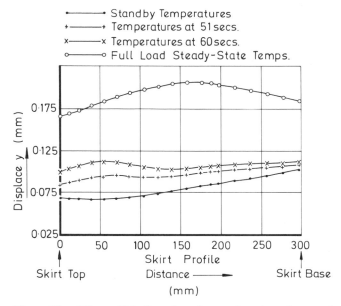

Figure 10   Skirt radial distortion plot on thrust/non-thrust axis from standby to full load steady-state conditions

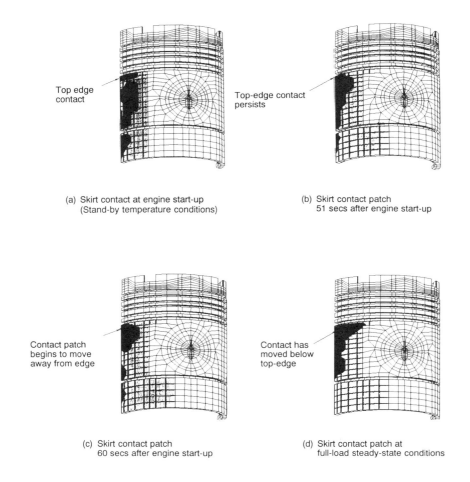

(a) Skirt contact at engine start-up
(Stand-by temperature conditions)

(b) Skirt contact patch
51 secs after engine start-up

(c) Skirt contact patch
60 secs after engine start-up

(d) Skirt contact patch at
full-load steady-state conditions

Figure11 Skirt contact patch comparison from standby to full-load steady-state conditions.

the anti-thrust side were highest at the top of the skirt during the early stages of starting. The increased compression pressures seen at the nuclear power plant during starting (Section 3), coupled with a lack the relieving effects of inertia at low speed, combined to produce this result. However, the peak contact pressure on the thrust side produced by firing loads is substantially higher than that seen on the anti-thrust side. Markings seen on pistons demonstrate that the contact pressures on the thrust side can be sustained whereas the contact on the anti-thrust side leads to very heavy bedding and wear. The explanation lies in the difference in lubrication conditions. On the thrust side high loads occur when the piston is on the down stroke and conditions are favourable for the formation of a hydrodynamic oil film. On the anti-thrust side the load is applied during the up-stroke, and is therefore applied via a boundary lubricated contact, which has a far lower load bearing capacity.

6. DISCUSSION

The crankcase explosions at the nuclear power plant were undoubtedly caused when skirt rubbing loads exceeded the capacity of the skirt lubrication at some time prior to failure. A reduction in the quality of the lubrication conditions, an increase in skirt loadings

or, a combination of both, could have caused the problem. Equally, there are many possible reasons why skirt loadings and/or lubrication conditions could be adversely affected. The data presented in this paper covers only part of the investigation carried out but it shows how a feature of the piston design was identified which would adversely affect skirt lubrication during start and loading transients.

From the motion analysis it can be seen that the piston is held against the anti-thrust side throughout the compression stroke when the compression pressures are increased by the turbocharger as load is increased or, by the operation of the start air system. When compression pressures are lower the piston moves away from the liner towards the end of its compression stroke thus relieving the loads on the anti-thrust side. The shape of the piston during the early stages of a rapid start and load is also not as favourable as it is at normal operating conditions. Therefore, it can clearly be seen that, due to the particular design of the piston, the operating conditions for the piston skirt are most arduous during the start and early part of a rapid loading. A slightly less rapid rate of loading allows the piston time to change to a more favourable shape before its motion becomes more arduous.

However, this finding could not be held solely responsible for the crankcase explosions as this type of piston has been employed for a number of years in many different applications with a very low reported incidence of seizures and/or crankcase explosions. The problem at this nuclear power plant was that a combination of circumstances conspired to reduce the quality of the lubrication conditions:

-   A lubricating oil was being used which, although normally perfectly satisfactory for this type of engine and application, nevertheless had a slightly lower extreme pressure rating than oils used at other nuclear power plants.
-   The monthly test procedure employed was severe, in terms of rate of loading after start, and was used more frequently than other power plants.
-   The particular starting air system configuration employed which enabled the engine to consistently meet the rapid starting requirement (start to synchronous speed in less than 10 seconds) caused an increase in compression pressures and, therefore, piston side loads in the first few cycles of start.

With the resulting reduction in quality of the lubrication conditions the engines were put at greater risk of scuffing than most of the other users of this type of engine and this could explain why the problem appeared in diesel engines serving as rapid start, emergency generator sets.

The electric utility have now amended the testing procedure of the engines and have also improved the quality of piston skirt lubrication by removing a lower oil control ring and piston pin end caps and changing to an oil with a higher extreme pressure rating. They are continuing to monitor the performance of the pistons and are also considering, for the long term, the adoption of a revised piston design.

7. CONCLUSION

The investigation involved the cooperative effort of at least five very diverse organisations, including the engine manufacturer, the electric utility, two engineering firms and a municipal electric plant. An engine at a rather remote site in Iowa was modified, instrumented and tested in under six months in order to support the continued operation of a nuclear power plant.

The work has demonstrated that a combination of advanced engine simulation and modelling techniques, validated by test work on a similar but not identical engine, can successfully identify the root cause of a piston problem that could not be investigated directly.

With the recent and continued advancement of modelling techniques coupled with the availability of ever more sophisticated but easy to use instrumentation, the designer will have the necessary tools to investigate potential problems with new component designs, particularly for those to be used in unusual or more arduous applications. The aim for the future must be to make full use of these tools to ensure development and testing time are reduced to a minimum.

8. ACKNOWLEDGMENTS

In preparing this paper, the authors would like to acknowledge the contribution of their colleagues and associates at the companies involved in the work. In particular:

W. Rhoades of Pennsylvania Power and Light Company,

J. J. Fowler, M. R. Stott, A. D. S. Bedi and R. Keribar of Ricardo International,

D. G. Evans, C. L. Haller and L. E. Lehman of MPR Associates, Inc.

and

W. Bohle, J. Duhrkopf, T. Duhrkopf, and R. Jergens of the Sumner Municipal light Plant, Sumner, Iowa.

The authors would also like to thank the management of the companies involved in the work described, for their permission to publish this paper.

REFERENCES

1. Csysz, F., "Crankcase Overpressurisation at Susquehanna", EPRI Seminar on Diesel Generator Operation, Maintenance and Testing, 1990.

2. Wibberley, D., "Ricardo CHASE Hardware Overview", Ricardo DP 88/1884

3. Smith, A. V. and Clarke, D. P., "Engine Component Optimisation - The Development of a Predictive CAE Based Model of Piston/Liner Interaction", FISITA 92, CIMAC Seminar - Computation and Simulation

4. Ricardo Publication "Computer-Aided Engineering Software for Engines and Propulsion Systems"

DEVELOPMENT AND FIELD TESTING OF A TWO-SPEED COGENERATION/ELECTRICAL PEAK SHAVING SYSTEM UTILIZING AUTOMOTIVE-DERIVATIVE GAS ENGINES

Luco DiNanno and Robert Raymond
Tecogen Incorporated
Waltham, Massachusetts

Allen Wells
Gas Research Institute
Chicago, Illinois

ABSTRACT

A dual-function engine-driven system which operates in two modes is described. Most of the system's operating life is in a relatively low speed mode where the cogeneration of heat and electricity is its function. For shorter periods, the system operates in a higher speed mode where its output is used to shave electrical demand peaks.

This paper describes the design and laboratory development of the Cogeneration/Peak Shaving System and includes the results to date of a field test which began in late 1991. Performance data are presented along with the evolution of the various subsystems.

Experience in the development and application of automotive gasoline engines for natural-gas-fueled commercial products is discussed with special reference to the unique requirements of the two-speed system.

INTRODUCTION

The Peak Shave/Base Load Cogeneration System is an outgrowth of Tecogen Inc.'s packaged cogeneration systems technology. In the smaller sizes (i.e., 30, 60, and 75 kW) these systems utilize an automotive-derivative natural gas engine rather than an industrial gas engine. In addition, induction generators are used to minimize switchgear costs, and the units are equipped with microprocessor controls and remote monitoring capability to improve maintenance economics.

At the 60-kW rating, a 454-cubic inch displacement (CID) engine manufactured by General Motors and modified by Tecogen has been shown to operate in excess of 20,000 hours with only routine maintenance. Its cost at that power rating, in dollars per kilowatt, is roughly one quarter that of an industrial engine.

With this very positive experience, it seemed likely that the 454-CID engine could operate at significantly higher power levels for shorter periods of time while still maintaining good operating economics. This conclusion led to the concept of a combined cogeneration and electrical peak shaving system, which would further enhance the overall economics of the system relative to a conventional cogeneration system and open up a broader market.

While there are a number of approaches that could be taken in exploiting the concept of a dual-power level system, a two-speed generator was adopted for this concept as being the most practical for a near-term, two-power-level product. A simple four-pole induction machine could be built with external connections which would allow either two-pole or four-pole (3600 rpm or 1800 rpm, respectively) operation. The engine(s) could be directly connected to the shaft of the generator, thereby eliminating the need for a clutch and gearbox.

The peak shave/base loaded cogeneration system selected for development consists of two 454-CID engines coupled to a single two-speed induction generator equipped with a double-ended shaft. Since the engine is available in either rotation, this arrangement poses no problem in modifying the engine. At the two operating conditions, 1800 and 3600 rpm, the goal is to have each engine operating at 80 kW and 160 kW, respectively.

There are two reasons for selecting an induction generator for this application. The first is low first cost. This, of course, determines that the unit cannot operate as a standby or emergency generator. The second reason is that a brief inquiry as to the feasibility of a two-speed synchronous generator showed that no such product is available and that a significant development effort would be required to produce such a machine.

At the present time, a laboratory version with a system rating of 160 kW at 1800 rpm and 320 kW at 3600 rpm has been built for development and testing at Tecogen. A second prototype field experiment unit has been built for Baltimore Gas and Electric and is being tested at the Marriott Hunt Valley Inn in the Baltimore area.

SYSTEM DESCRIPTION

An overall system schematic for the naturally aspirated version is shown in Figure 1. Here the double lines indicate exhaust gas plumbing, the heavier single line shows lubricating oil system plumbing, and the lighter single line shows coolant system plumbing. The two engines and generator form a rigid beam that can also be seen in Figure 2, a photograph showing the side view of the field experiment unit. An end view of the unit is shown in Figure 3. The engines are equipped with water-cooled exhaust manifolds to improve heat collection efficiency and minimize thermal radiation to the surroundings. All of the heat collected from the system is rejected to the load heat exchanger, which is part of the packaged system. The user's thermal load sees only the colder side of this heat exchanger.

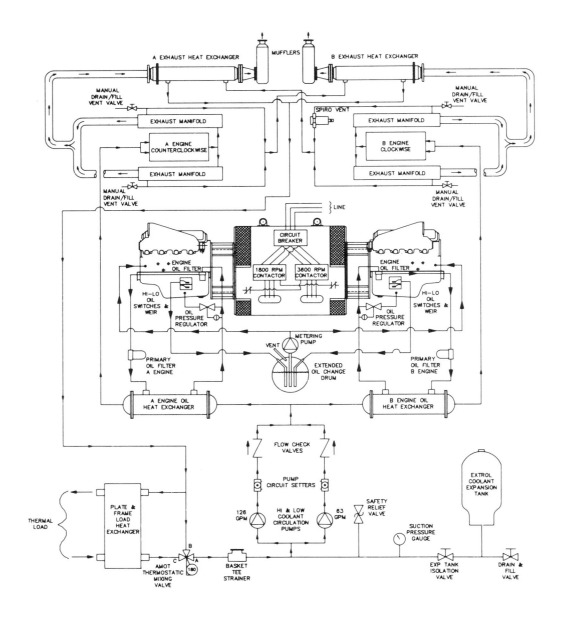

FIGURE 1. PEAK SHAVE/BASE LOAD COGENERATION SYSTEM SCHEMATIC

FIGURE 2. SIDE VIEW, PEAK SHAVE/BASE LOAD COGENERATION SYSTEM

FIGURE 3. END VIEW, PEAK SHAVE/BASE LOAD COGENERATION SYSTEM

As Figure 1 shows, heat rejected by the engine lube oil system, jacket, exhaust manifolds, and exhaust heat exchanger is absorbed by the coolant and delivered to the load heat exchanger and a thermostatic mixing valve which controls the temperature of the coolant returning to the system.

Most of the thermal load side of this system is like any other cogeneration system, but is different in one significant respect. Because the engines operate at two very different loads (and speeds), the heat rejection of the engines at the two conditions is quite different. A motor-driven pump circulates coolant through the entire system. It is desirable to keep the coolant temperature into the engines and the temperature rise across them relatively constant for reasons of engine durability. To accomplish this, two different flow rates are provided by two main circulating pumps.

The lubricating oil system is unique in several ways. The oil change interval is extended by the 55-gallon reservoir, which recirculates oil with the main engine sumps at the rate of about one gallon per hour. This system, together with the engine filters and a separate, very large main oil filter for each engine, should allow at least 2,000 hours and up to 4,000 hours between oil changes, depending on the amount of time the unit operates in the peak shaving mode.

The module containing all of the plumbing and heat exchangers can be seen under the engine and generator assembly (Figures 2 and 3). The heat exchangers provide longitudinal stability to the module, which consists of 4 ribs spaced one under each engine and two under each end of the generator. This construction technique eliminates the use of a heavy subframe channel or I-beams, which are basically unnecessary since the engine-generator assembly is already a very stiff beam.

The engine-generator assembly is mounted to the tops of the subassembly ribs through vibration isolators. Connection to the plumbing from the engine is through easily disassembled fittings. The electrical control box (shown in Figures 4 and 5) is mounted to the assembly and contains the control system and all of the switchgear and other electrical equipment. The control system of the peak shave/base load system uses a new single-board controller hardware platform based on a Motorola 68000 series chip which is being introduced across the Tecogen product line. The control system shown mounted to the electrical control box door in Figure 5 incorporates a complete operating system and is programmed to operate the peak shave/base load system automatically, including the engine speed switching function. The controls also monitor the operation of the system and store operating data for later retrieval and analysis. The control system permits either on-site or remote operation of this machine as well as data retrieval and provides the means for remote dispatching of peak shaving capability.

FIGURE 4. ELECTRICAL CONTROL BOX, PEAK SHAVE/BASE LOAD COGENERATION SYSTEM

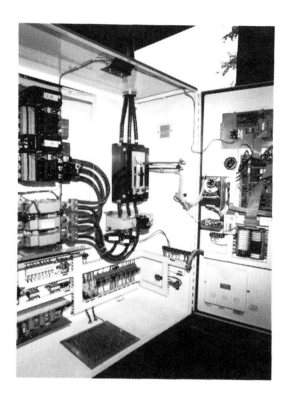

FIGURE 5. VIEW OF CONTROL SYSTEM AND SWITCHGEAR IN ELECTRICAL CONTROL BOX,
PEAK SHAVE/BASE LOAD COGENERATION SYSTEM

There are basic differences between the peak shave system control logic and the typical logic of other Tecogen cogeneration systems that are programmed into our controller. These are:

1) After the start mode is completed and the unit is "running," a check must be performed to ensure that <u>both</u> engines are operating. A comparison of actual idle speed to an expected idle speed with both engines operating is made by the controller; lower than expected idle speed results in shutdown, and a starting error message is displayed.

2) A means of equalizing the power output of the two engines is required. The approach adopted is to use a stepper motor actuator at each engine and to operate them in tandem during transient conditions. At full load at either speed, the controller reverts to independent control of the two throttle stepper motors, with one becoming the "lead" engine attempting to hold full power and the second becoming the "follower" engine attempting to reduce the difference in manifold pressure between itself and the lead engine to zero.

3) A means of changing the operating mode while the unit is running (as opposed to shutting down and restarting in the new mode) is required in this system. The possibilities are:

   a) Shutdown (no heat or peak shaving required).
   b) Go to peak shave mode.
   c) Go to cogeneration mode.

This situation is easily adapted into our controller. The new logic has been developed and demonstrated.

The entire unit may be lifted with a forklift truck from underneath or by an overhead crane through the generator. Table 1 gives the preliminary specifications for the naturally aspirated version of the Peak Shave/Base Load Cogeneration system.

TABLE 1

TECOGEN PEAK SHAVE/BASE LOADED COGENERATION (PS/BLC) SYSTEM WITH NATURALLY ASPIRATED ENGINES

Preliminary Specifications

Output:
- Cogeneration Mode at Engine Speed of 1800 rpm
  - Electrical — 160 kW, 460 V, 3 Ph, 60 Hz, Power Factor 0.92
  - Thermal — 1,000,000 Btu/hr Hot Water

- Peak Shave Mode at Engine Speed of 3600 rpm
  - Electrical — 320 kW, 460 V, 3 Ph, 60 Hz, Power Factor 0.94
  - Thermal — 2,100,000 Btu/hr Hot Water*

Input:
- Cogeneration Mode at Engine Speed of 1800 rpm
  - 1830 scfh Natural Gas (1020 Btu/scf HHV)
  - Coolant Pump Motor — 1.2 kW, 460 V, 3 Ph, 60 Hz
  - Controls and Misc. — 1 kW, 460 V, 3 Ph, 60 Hz

- Peak Shave Mode at Engine Speed of 3600 rpm
  - 4130 scfh Natural Gas (1020 Btu/scf HHV)
  - Coolant Pump Motor — 4.7 kW, 460 V, 3 Ph, 60 Hz
  - Controls and Misc. — 1 kW, 460 V, 3 Ph, 60 Hz

Efficiency:
- Cogeneration Mode
  - Electrical — 29.2%
  - Combined Electrical and Thermal — 82.8%

- Peak Shave Mode
  - Electrical — 26%

Controls:
- Microprocessor based with Tecogen Single-Board Controller. Fully automatic startup, monitoring, load following, dual-speed, and power operation and shutdown. Programmable set points. Digital display. Optional remote monitoring.

Dimensions: 12' long x 7'6" wide x 6'6" high

Weight: 10,000 lb approximately

*Based on load return water temperature of 170°F. Specifications subject to change without notice.

SYSTEM FIELD TEST

The unit shown in Figures 2 through 5 was purchased by Baltimore Gas & Electric (BG&E) and installed at The Hunt Valley Inn (Marriott) in late November 1991. The system supplements the Inn's boilers in providing domestic hot water and space heating. BG&E's interest in the system derives from their interest in Demand Side Management as a means of meeting their future capacity requirements. The operational concepts being demonstrated in their field test are as follows:

- Utility ownership
- Thermal output discount to the host site
- Utility dispatch to peak shaving mode during periods of high system electrical demand

This arrangement represents the most favorable duty cycle since the number of hours the unit would have to operate in the electrical peak shaving mode is far fewer than if the Hunt Valley Inn were trying to shave its own peak electrical demand charges.

The system shown in Figure 1 is connected to intermediate heat exchangers which transfer its heat to the Inn's domestic hot water and space heating storage tanks. It is also connected to a dump radiator so that it can be run in the peak shaving mode when there is no requirement for its thermal output.

Since the unit is a field experiment, it is completely instrumented with electrical, thermal, and gas consumption meters. In addition, the microprocessor is connected to a data acquisition system which stores unit operating pressures and temperatures and records alarms which cause unscheduled shutdowns. From this information, a complete record of unit output and availability is maintained.

Figure 6 shows the unit's availability and utilization from start-up through April 1992. Table 2 lists the alarm summary through the same period. The low availability during January, February, and March was due to engine failures. In addition to serving as a field test of the Peak Shave/Base Load Cogeneration system, this unit also served as a test bed for developing high-compression-ratio pistons for the 454-CID engine. This effort is described in more detail in a following section. The first engine failure shown in Table 2 (1/8/92) was due to a scuffed piston. That engine was rebuilt in place at the site with minor honing of the scuffed cylinder. The remaining engine failures listed in Table 2 were due to a broken crankshaft (2/16/92) and a cracked piston (3/6/92). The crankshaft was cast iron and failed in bending at a fillet. Replacement engines have been equipped with standard forged cranks. Piston cracking is discussed more extensively in a following section. These pistons gave a compression ratio of 12.2 to 1.

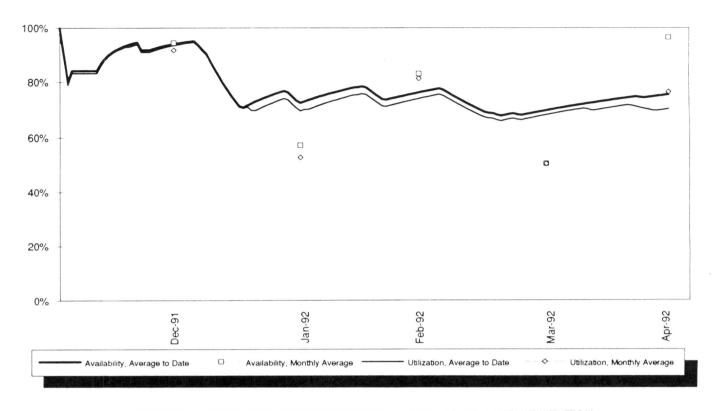

FIGURE 6. BG&E FIELD EXPERIMENT UNIT – AVAILABILITY AND UTILIZATION

TABLE 2

BG&E FIELD EXPERIMENT UNIT (FEU)
ALARM SUMMARY REPORT

Alarm	Date	Comment	Number to Date
Coolant Pressure	12/4/91	Improper machine setup	1
None, System Shut Down	12/16/91	Electrical feeder problem	1
None, System Shut Down	12/17/91	Building distribution problem	2
None, System Shut Down	12/17/91	Dump loop recirculation pump problem	1
High Enclosure Temp	12/18/91	Enclosure fan failure	1
Cranking Failure	1/6/92	Starter failure	1
None, System Shut Down Manually	1/8/92	Engine failure	1
High Enclosure Temp	1/23/92	Thermostat mis-adjusted	2
Cranking Failure	1/28/92	Starter bolt failure	1
Coolant Pressure	2/1/92	Loop pump off (also caused 132 starts)	2
RMCS Shutdown	2/13/92	Site owner reported leak (very minor)	1
RMCS Shutdown	2/14/92	Shutdown to repair leak	2
Low Oil Pressure, Engine B	2/16/92	Engine failure (several related alarms)	1
Oil Level	3/6/92	Engine failure (several related alarms)	1
Overspeed Board Trip	3/18/92	Set improperly during service	1
Current Imbalance	3/18/92	Software bug	1
High Coolant Temp	3/18/92	Unexplained phenomenon in peak shave	1
Current Imbalance	3/18/92	Software bug	2
Oil Level (multiple occurrences)	3/19/92	Unexplained phenomenon in peak shave	2
Oil Level	3/23/92	Unexplained phenomenon in peak shave	3
Oil Level	3/24/92	Unexplained phenomenon in peak shave	4
High Oil Temp	4/8/92	Unexplained phenomenon in peak shave	1
None, System Shut Down	4/11/92	Building problem caused SBC to trip	1
Low Oil Pressure, Engine B	4/14/92	Building breaker (several related alarms)	1
Low Oil Pressure, Engine B	4/20/92	Building distribution problem	2
None, System Shut Down	4/20/92	Building distribution problem	3
High Oil Temp	4/22/92	During service (several occurrences)	2
Oil Level	4/23/92	During service (several occurrences)	5

Note: Shaded alarms are due to unit.

133

Problems with the lube oil system of one engine are also noted in Table 2. This engine was found to have a worn oil pump drive shaft, which may have reduced oil flow to the cooler enough to overheat the oil.

The wider difference between availability and utilization shown in Figure 6 for April reflects a lower requirement for space heat. Figure 7 shows a breakdown of the various modes for April. The causes for system forced outages are shown in Table 2 as the unshaded areas in April. All of these are related to the electrical system in the hotel.

In the cogeneration mode the unit responds to the building thermal requirements in either the domestic hot water or the heating system storage tanks. Peak shaving is controlled remotely by BG&E, for test purposes, and is automatically set for four hours a day — two in the morning and two in the afternoon — five days per week.

This is probably more peak running than a single system in a fully deployed situation would have to run, but it allows BG&E to determine the probability that any single unit will be available when called upon to do peak shaving.

SYSTEM DEVELOPMENT

The system shown schematically in Figure 1 and in photographs in Figures 2 through 5 was evolved from a laboratory prototype. This prototype is shown schematically in Figure 8. A comparison of Figures 8 and 1 shows significant differences in the coolant, lube oil, and exhaust systems of the lab prototype and field test unit. This section will detail how these subsystems evolved.

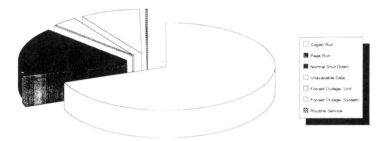

FIGURE 7. BG&E FIELD EXPERIMENT UNIT – TIME BREAKDOWN FOR APRIL 1992

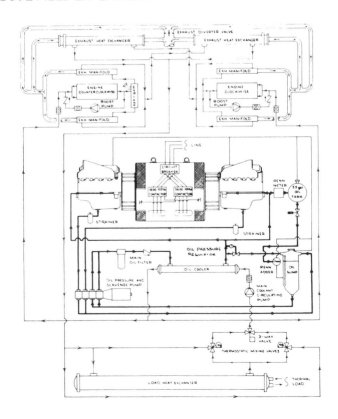

FIGURE 8. PROTOTYPE PEAK SHAVE/BASE LOAD SYSTEM SCHEMATIC

Coolant System

The key requirement of the cooling system is that it provide the proper engine temperature level and temperature rise at the two conditions at which it is required to operate. There are a number of ways of doing this while minimizing the parasitic power required to operate the circulating pumps. There are enough additional components in the system (oil cooler, load heat exchanger, exhaust manifolds, and exhaust heat exchangers) that the increased pressure drop at the required flow rate rules out the use of the stock engine driven pumps. Maintenance problems with belts also favor the use of motor driven pumps.

The system shown in Figure 8 had a single main circulating pump which operated in both the cogeneration and peak shaving mode. Two boost pumps, one on each engine, increased the flow rate through the engine during peak shaving to keep the temperature rise at 15°F. Selecting a lower set point thermostatic valve at the suction side of the main circulating pump allowed cooler coolant to be supplied at the mixing point between recirculated coolant and coolant from the load heat exchanger. This system operated well in the laboratory, but the equipment cost was high. During testing it was found that the pressure drop through the system was less than calculated in the design phase, thereby reducing the parasitic power requirement.

In view of these results, a trade-off study of various cooling systems was carried out, and the results are shown in Table 3. On the basis of this study, the twin main pump approach of Figure 1 was selected as the optimum approach, and the first field test unit was built with this system. Subsequently, a pump was found with the proper characteristics to give the head and flow required at pump speeds of 1800 and 3600 rpm. Driving this pump with a two-speed motor gives a further equipment cost advantage over the two-pump approach of Figure 1.

Lube Oil System

With the lube oil system, the basic requirement is to keep the sump temperature at a level which gives acceptable oil life. This is accomplished in the system shown in Figure 1 by circulating lube oil through individual coolers. A high-capacity lube oil pump is fitted to the engine with its internal bypass shimmed closed. This ensures that the full flow capacity of the pump is delivered to the cooler, after which an external relief valve supplies pressurized oil to the bearings and the excess capacity is relieved to the sump.

In the naturally aspirated version of the system, the arrangement shown in Figure 1 is sufficient to keep the sump temperature in the 215°F range in the peak shaving mode. The dry sump configuration of Figure 8, with separate motor-driven scavenge and pressure pumps, was conceived in anticipation of a turbocharged version of the system where 300 bhp per engine would be developed. At that power level some additional oil flow for piston cooling was also anticipated. Under these conditions, the engine driven pump could not deliver sufficient oil flow to the cooler to keep the sump at 215°F.

The system shown in Figure 8 worked well but was unnecessarily expensive for the naturally aspirated version of the system. An intermediate system between those shown in Figures 1 and 8 was also tested. This system utilized the arrangement shown in Figure 1 but had a single heat exchanger and relief valve. This arrangement made it difficult to balance the oil levels in the two engines, so the completely separate arrangement shown in Figure 1 was adopted.

The capacity of the sumps is extended by 55 gallons with an external drum. One gallon an hour is pumped from the drum into the sumps of each engine, and an equal amount (less consumption) is allowed to drain over a weir attached to the oil pan and back to the drum. This arrangement allows up to 4,000 hours between oil changes.

TABLE 3

INITIAL COST VERSUS OPERATING PAYBACK FOR VARIOUS COOLANT SYSTEMS
BASIS – 170°F RETURN WATER TEMPERATURE
6,000 HOURS COGENERATION AND 1,000 HOURS PEAK SHAVING/YEAR
$0.05/KW-HR ELECTRICITY COST

Pump System	Single Main Pump		Throttled Single Main Pumps		Twin Main Pumps		Single Main and Boost Pumps	
	Cogen	Peak Shave	Cogen	Peak Shave	Cogen	Peak Shave	Cogen	Peak Shave
Annual Operating Cost (Incremental) ($)	Base		−380	+26	−1522	0	−1455	−157
Incremental Equipment Cost ($)	Base		+342		+595		+1562	

Exhaust System

With the exhaust system, the trick is to balance performance with cost and reliability. The prototype design of Figure 8 incorporated an exhaust bypass valve for the peak shaving mode. This would allow some exhaust to bypass the exhaust heat exchangers, thereby reducing the size of the dump heat radiator if the heat was not required during peak shaving.

During development, this bypass valve proved to be a problem. Corrosion and leaking were the major problems, and there were some actuator difficulties as well. It was determined that eliminating the bypass valve and associated plumbing and re-sizing the exhaust heat exchangers to reduce pressure drop and heat transfer in the peak shave mode constituted a more cost-effective solution in terms of first cost and maintenance cost vs. some small loss in heat recovery in the cogeneration mode.

The development effort expended on the laboratory prototype results in a $7,200 material savings, 25 percent less labor, and improvements in cost of operation and reliability.

Engine Considerations

This portion of the paper will summarize the status from the point of view of performance and durability of the 454-CID engine used in the Peak Shave/Base Load Cogeneration System described in the previous sections as well as other Tecogen Inc. engine-driven products. This engine is produced by the Powertrain Division of General Motors at their plant in Tonawanda, NY, USA, and is modified and converted by Tecogen Inc. to burn natural gas. A program to improve the performance and maintenance requirements of this engine was undertaken by Tecogen Inc. with assistance from the Gas Research Institute and General Motors. We will outline the results of the naturally aspirated performance improvement portion of this program which consisted of re-optimization of the camshaft for the (relatively) low speed cogeneration application and development of a high-compression-ratio piston to take advantage of the high effective octane number of natural gas.

Three pistons were evaluated as part of the performance improvement phase of this program. They were the stock piston, which has been part of the Mark IV build of the 454-CID engine for a number of years; an aftermarket piston from TRW with a dome, which gave a compression ratio of 12.2 to 1; and a prototype hypereutectic piston from Zollner developed by GM and Zollner especially for this program. Some characteristics of these pistons are shown in Figure 9.

As Figure 9 indicates, the dome on the hypereutectic piston was constructed so that it could be progressively machined to give any desired compression ratio between 9.3 and 12.2 with the 366 heads. Thus, it was possible to test an intermediate compression ratio without changing the cylinder head configuration (or the pistons and rings).

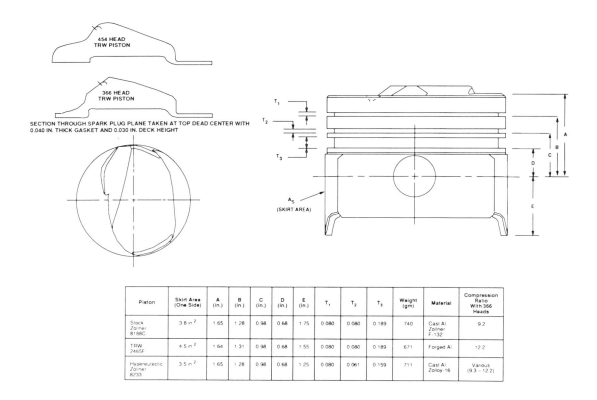

Piston	Skirt Area (One Side)	A (in.)	B (in.)	C (in.)	D (in.)	E (in.)	T_1	T_2	T_3	Weight (gm)	Material	Compression Ratio With 366 Heads
Stock Zollner 8188C	3.8 in²	1.65	1.28	0.98	0.68	1.75	0.080	0.080	0.189	740	Cast Al. Zollner F-132	9.2
TRW 2465F	4.5 in²	1.64	1.31	0.98	0.68	1.55	0.080	0.080	0.189	671	Forged Al.	12.2
Hypereutectic Zollner 8233	3.5 in²	1.65	1.28	0.98	0.68	1.25	0.080	0.061	0.159	711	Cast Al. Zolloy-16	Various (9.3 - 12.2)

FIGURE 9. PISTON CHARACTERISTICS

Performance

Figure 10 shows the results of testing the three pistons at various compression ratios with a stock GM cam (3545) and two experimental cams.

As Figure 10 indicates, the performance of the hypereutectic piston at three compression ratios follows more closely the analytical simulation supplied by GM than did results with the stock and TRW pistons. This is probably due to the fact that all three compression ratio tests were carried out with the same pistons, rings, and cylinder heads – something that was not possible to do with the TRW piston.

Again referring to Figure 10, if the difference in horsepower (or BMEP) between the stock piston at 9.2 to 1 and the hypereutectic piston at 9.3 to 1 (both with the TE-100 cam) is attributed to a difference in friction mean effective pressure (FMEP), then the improvement in BSFC shown is consistent with that assumption. The difference would be about 3 psi at 1800 rpm or 20 psi FMEP for the stock piston and 17 psi for the hypereutectic.

The data of Figure 10 are at a nominal manifold pressure of 27 in. Hg abs and a fuel/air ratio which gave 2 percent O_2 in the exhaust. At wide open throttle, a corrected BMEP of 119 psi was achieved with a measured brake efficiency of 32.3 percent; all with the hypereutectic piston and TE-100 cam.

At the 12.2 to 1 compression ratio with pipeline quality natural gas, the spark could typically be advanced 10 to 15 degrees beyond best power before detonation was encountered at 1800 rpm and a fuel/air ratio giving 2 percent O_2 in the exhaust.

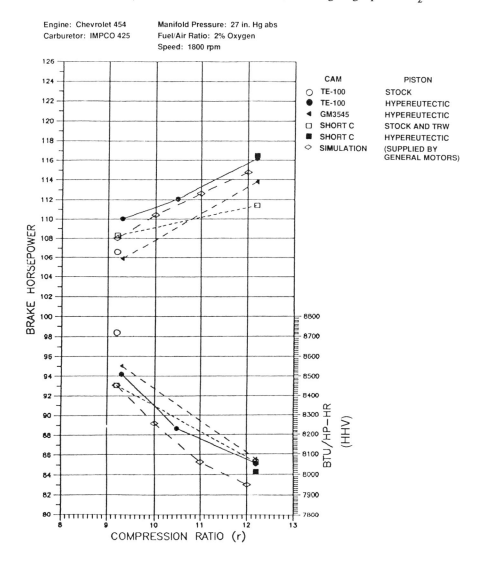

FIGURE 10. BHP AND BSFC VS. COMPRESSION RATIO FOR VARIOUS CAMS AND PISTON TYPES

A special slow-speed cam had been designed by Chevrolet for use in the natural gas engine. This cam was designated Short C, and its characteristics are listed in Table 4. Subsequent endurance testing of this cam showed that the relatively high design stress (see Table 4) resulted in excessive wear after 5,000 hours of operation. The cam profile was then modified to give a new cam designated as TE-100. The characteristics of the TE-100 are given in Table 4 together with the production cam currently supplied with the 454-CID engine.

The design philosophy for the TE-100 profile was to use positive acceleration values significantly higher than those used with the Short C since the maximum stress occurs at the nose of the cam during periods when the cam is under negative acceleration and at slow speed. This technique allowed somewhat higher lift than was achieved with the Short C while maintaining a lower stress level.

Durability

We will attempt to summarize our experience with the 454-CID over the past ten years with emphasis on those issues of wear and reliability which affect the Peak Shave/Base Load Cogeneration System.

To do this, it is convenient to consider reliability and durability as being influenced by normal wear and premature failure.

Experience accumulated with this engine in one type of duty cycle, i.e., constant 1800 rpm and 84-psi BMEP cogeneration service, allowed us to demonstrate that the observed life (15,000 to 20,000 hours) could be correlated with exhaust valve recession rate and piston ring wear rate.

In cases where the engines do not "wear out" and there are premature failures, these can be attributed to one of the following three categories:

1. System induced engine failures, e.g., repeated overheating of the engine due to a poorly designed cogeneration system, or hydrolocking of the engine due to exhaust heat exchanger leaks.

2. Quality control in engine manufacture.

3. Engine design and materials specification changes.

Due to the low cost of the engine and the ease with which it can be replaced, normal wear is not a significant problem in current Tecogen applications. The severity of premature failure problems is periodic, with one of the above three categories usually overshadowing the other two at any given time. A detailed examination of our service records has shown that premature engine failures are system-induced by about a ten to one margin over quality defects in the engine.

TABLE 4

CAMSHAFT CHARACTERISTICS
(Rocker Arm Ratio − 1.7)

All Events at 0.005" Cam Lift		Short C	TE-100	3545	6209
Exhaust Opens (° Before Bottom Center)		54°	57°	63°	75°
Exhaust Closes (° After Top Center)		8°	5°	19°	24°
Intake Opens (° Before Top Center)		16°	23°	15°	32°
Intake Closes (° After Bottom Center)		49°	39°	53°	87°
Overlap (°)		24°	28°	34°	56°
Total Lobe Lift (in.)	Intake	0.235	0.257	0.235	0.297
	Exhaust	0.235	0.257	0.253	0.297
Stress (Hertz) (psi)		152,900	137,976	135,587 (Exh. Lobe)	——

Normal Wear

Figure 11 is a plot of BMEP vs. piston speed with lines of constant brake horsepower per square inch of piston area. The closed symbols represent the continuous ratings of various industrial (diesel-derivative) gas engines while the open figures represent the ratings of various automotive-derivative gas engines. The open figures with numbers represent operating conditions in which Tecogen has collected or is presently collecting wear data on the 454-CID engine. Some of these also correspond to current or planned Tecogen products, as follows:

[1] Represents the running condition for the CM-60 cogeneration module, with about 400 units in service. This represents the condition where 15,000-hour to 20,000-hour head and block life is achieved in good installations. The compression ratio is 9.2 to 1 using the stock piston of Table 4.

[2] Represents the running condition for the CM-75 cogeneration module, which is currently achieved by replacing the stock GM cam with the TE-100 of Table 4. A high-compression-ratio piston is being developed to allow this rating with more margin and higher efficiency. It is also being developed for use in the Peak Shave/Base Load Cogeneration System in base load operation. Field tests of up to 8100 hours have been achieved, but piston durability is still a problem and will be discussed more fully.

[3] Represents the maximum output of the 150-ton chiller drive engine. One engine was tested at Southwest Research Institute for 5,000 hours. This system operates along the line designated as "6" in Figure 1. Approximately 150 units are in the field.

[4] Represents a test point where four engines were endurance-tested at Southwest Research Institute for up to 3,000 hours.

[5] Represents the operating condition of the proposed turbocharged version of the peak shave/base load system in the peak shave mode (base load mode is [2]). A number of engines have been endurance tested at this condition (piston durability and intake valve recession have been the major problem areas). Up to 2000 hours per year are required at this condition.

[6] See under [3].

[7] Represents the operating condition for the naturally aspirated, high-compression-ratio version of the peak shave/base load system as described in this paper.

[8] Represents the maximum power requirement of the 500-ton chiller drive engine. Wear data are being collected from laboratory and field test units.

[9] Represents the maximum operating envelope of the 500-ton chiller drive.

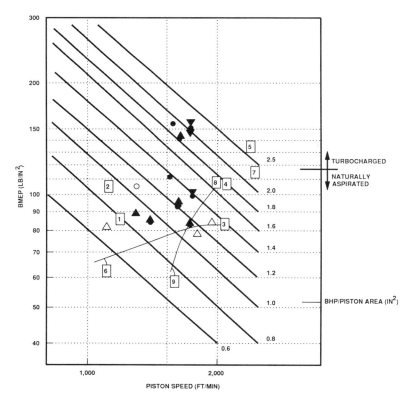

FIGURE 11. BRAKE MEAN EFFECTIVE PRESSURE VS. PISTON SPEED

Figure 12 shows the hours to exceed service specifications vs. component wear rate for the 454-CID engine. These come from GM service manuals with the exception of valve recession, which is based on Tecogen experience and which assumes that valve recession is about equally divided between valve face wear and seat wear. Figure 13 shows a typical exhaust valve and seat wear pattern.

With the exception of valve recession, an engine will typically continue to operate satisfactorily after the service specifications of the components shown in Figure 12 are exceeded. However, the second compression ring is tapered, as shown in Figure 14, and this taper is important in oil consumption control. When the ring wears to the point that the taper is gone, oil consumption becomes excessive, and it is more cost-effective (from the point of view of service) to exchange the engine than to visit the site at too-frequent intervals for adding oil. It turns out that when this ring gap exceeds the service specification of Figure 12, the taper is gone.

What all of this means is that the life of the 454-CID engine can be predicted from Figure 12 if one knows what the wear rates of these components are at any particular operating condition. To illustrate how this works, Figure 15 shows valve recession rates for the various conditions of Figure 11. Condition [1] has an average recession rate of about 1 mil per 100 hours which, according to Figure 11, gives the 15,000 hour head life we typically observe.

Figures 16 and 17 show valve recession and second compression ring data taken at condition [2] of Figure 11 with 12.2 to 1 compression ratio. This is the base load rating of the Peak Shave/Base Load System. These data indicate that cylinder head life at the base load rating should be about 10,000 hours while the second compression ring (and hence oil consumption) should remain acceptable to 20,000 hours of operation.

The wear rates in the peak shave mode are, of course, substantially higher and data obtained during the field test described here will indicate their impact on engine life.

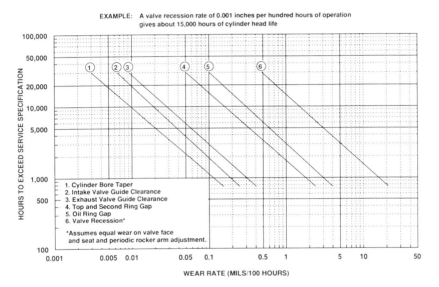

FIGURE 12. HOURS TO EXCEED SERVICE SPECIFICATIONS VS. WEAR RATE FOR VARIOUS ENGINE COMPONENTS OF 454-CID ENGINE

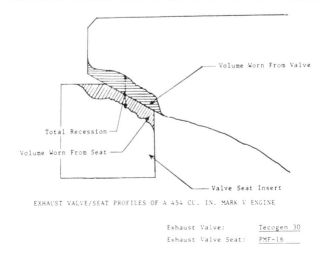

FIGURE 13. EXHAUST VALVE/SEAT PROFILES OF A 454-CUBIC-INCH MARK V ENGINE

140

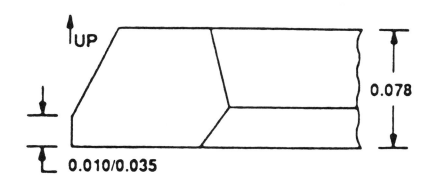

FIGURE 14. TAPERED SECOND COMPRESSION RING

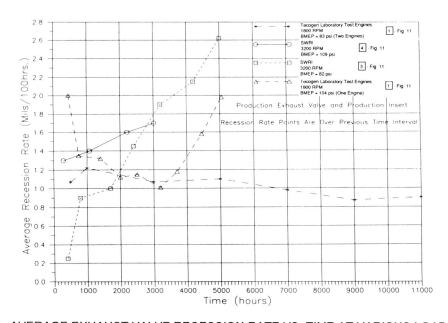

FIGURE 15. AVERAGE EXHAUST VALVE RECESSION RATE VS. TIME AT VARIOUS LOADS AND SPEEDS

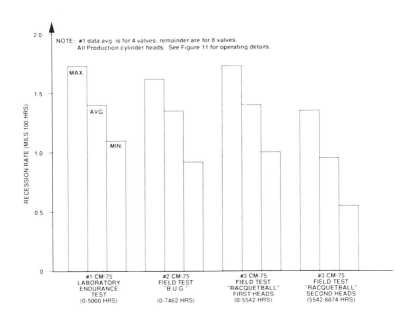

FIGURE 16. EXHAUST VALVE RECESSION RATE FOR THREE CM-75 ENGINES

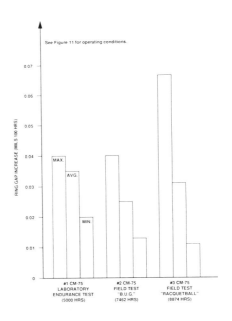

FIGURE 17. SECOND COMPRESSION RING WEAR RATE FOR THREE CM-75 ENGINES

Durability Issues

While satisfactory performance has been obtained from the high-compression-ratio piston (see Figure 10) the life of the cast hypereutectic alloy piston has been limited. The best life obtained to date at condition [2] of Figure 11 is 8100 hours. A number of aftermarket forged aluminum alloy pistons have also been tested at this condition and have proven to be somewhat less durable.

The failure mode is always the same, regardless of design or material; namely cracking on the loaded size of the pin bore. Figure 18 shows a typical failure and Figure 19 is a detail of the initiation site, a subsurface fatigue crack at an angle of 45° to the surface of the pin bore. This is indicative of high compressive stress.

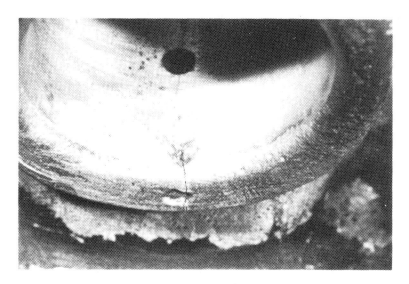

FIGURE 18. PIN BORE CRACK LOCATED AT THE TOP OF THE PIN HOLE SHOWING SECTION FRACTURED OUT OF THE SURFACE ADJACENT TO THE CAST RADIUS

FIGURE 19. FATIGUE CRACK INITIATION SITE
(Photos courtesy of Zollner Corporation)

Pulsator testing by the piston manufacturer indicates the hypereutectic piston should withstand firing pressures of 1050 psi, while an independent analysis by Perkins Technology Limited indicated 1,000-psi firing pressure should be sustainable. We have measured firing pressures, over limited time spans, which average at about 800 psi and have peaks slightly under 1,000 psi due to cycle-to-cycle variation.

The cast hypereutectic piston was modified to give a fourth generation version, as shown in Figure 20. Variations on this as shown in Figure 21 are currently under evaluation at conditions ②, ⑧, ⑦, and ⑤ of Figure 11. The stock configuration is a fixed

wrist pin pressed into the connecting rod. Provisions for a floating wrist pin have been made based on the assumption that at least some of the compressive loading may be due to heat generated by friction between the pin and pin boss. A floating pin allows some of the relative motion to occur between pin and connecting rod.

We have not attempted to either increase the pin bore diameter or the loaded side bearing area by use of a tapered connecting rod small end. These types of changes imply changes in the basic tooling that manufactures these engines.

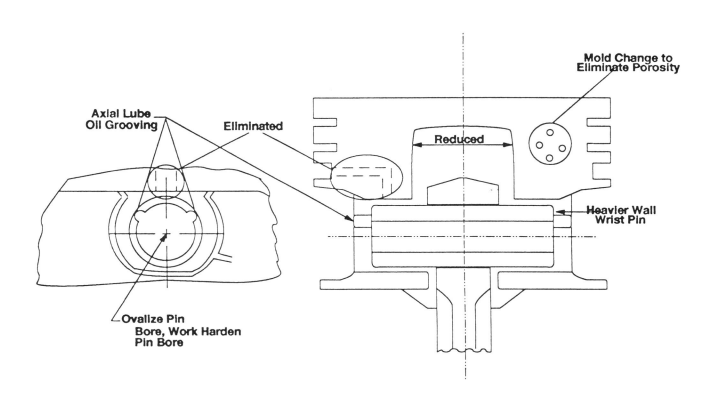

FIGURE 20. PISTON DESIGN CHANGES, 3RD TO 4TH GENERATION ZOLLNER CAST HYPEREUTECTIC

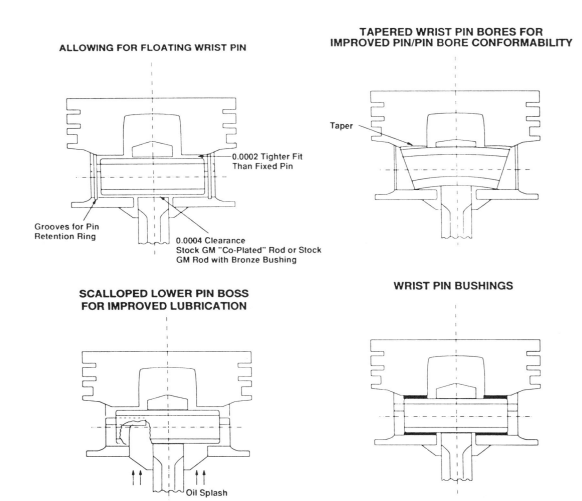

ALLOWING FOR FLOATING WRIST PIN

0.0002 Tighter Fit Than Fixed Pin

Grooves for Pin Retention Ring

0.0004 Clearance Stock GM "Co-Plated" Rod or Stock GM Rod with Bronze Bushing

TAPERED WRIST PIN BORES FOR IMPROVED PIN/PIN BORE CONFORMABILITY

Taper

SCALLOPED LOWER PIN BOSS FOR IMPROVED LUBRICATION

Oil Splash

WRIST PIN BUSHINGS

FIGURE 21. PISTON TEST PROGRAM

SUMMARY

This paper has described the latest application for automotive-derivative gas engines, the two-speed Peak Shave/Base Load Cogeneration System. While the economics of this system are enhanced by the use of 12.2 to 1 compression ratio pistons, the standard 9.2 to 1 engine already proven in cogeneration and chiller drive service represents a still attractive alternative while development continues on the advanced piston.

A little further in the future, we look forward to a turbocharged version of this novel system.

CONTINUOUS FUEL EFFICIENCY MONITORING

Daniel C. Zebelean
Kern River Gas Transmission
Salt Lake City, Utah

Christine Snitkin
Northwest Pipeline Corporation
Salt Lake City, Utah

ABSTRACT

Northwest Pipeline (NWP) has developed and employed a fuel efficiency monitoring program to track differences between engine manufacturers' predicted fuel consumption and actual engine fuel consumption.

The data is obtained through a SCADA system. A computer program was created where equations are written and stored to represent the predicted fuel curves at various loads. Based on the operating data, the predicted fuel use is interpolated from these equations.

The purpose of this program is to locate sources of fuel inefficiency in reciprocating engines. Mechanical problems, incorrect computer software, and out-of-tolerance operating parameters are thereby discovered and resolved. This allows for increased operating efficiency and lower maintenance costs.

BACKGROUND

NWP consumed approximately 9,600,000 MMbtu's in reciprocating engine fuel gas in 1990. Fuel gas has always been measured and accounted for; however, the thermal efficiency of the fuel gas consumed was never quantified. Engineering management set a goal to audit the fuel consumption using engine manufacturers' predicted fuel consumption curves.

A report was created which tracks each engine's fuel efficiency. Unbalanced engines, faulty spark plugs, engines requiring overhauls, fuel valves, and even computer software and hardware problems were found based on this report. The report was utilized in correcting these discrepancies, and in doing so, the monthly heat rate for the reciprocating units went from 7957 to 7707 btu/bhp-hr. In monetary terms, this equates to a savings of $156,000 per year.[1]

DATA ACQUISITION

NWP's Supervisory Control and Data Acquisition (SCADA) system logs the necessary operating data required to retract the predicted fuel consumption from the manufacturers' fuel consumption curves. The SCADA system logs the engine torque and RPM data and stores this information in a data base. Knowing engine torque and RPM, the predicted fuel consumption can be determined from manufacturers' curves.

VARYING SPEED AND TORQUE

Engine manufacturers publish predicted fuel consumption data at 100% speeds and 100% torques. A procedure needed to be derived to predict fuel consumption at reduced speeds and reduced torques.

[1] Assuming operating 50,000 bhp (1/3 of NWP's reciprocating horsepower), 365 days per year, 1050 btu/ft^3, $1.50/MCF.

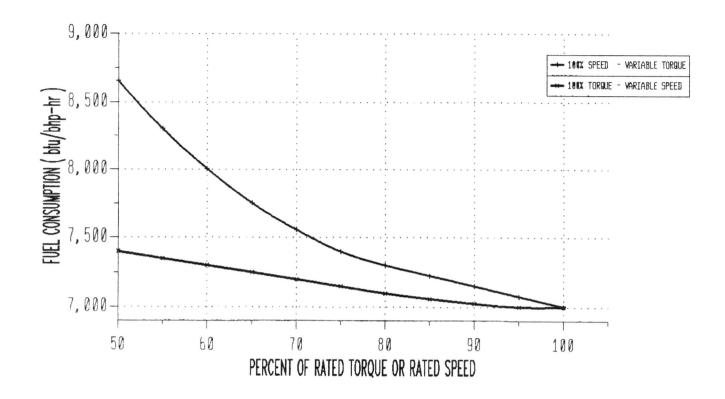

FIG #1 - GENERIC ENGINE FUEL CONSUMPTION GRAPH

Fig. #1 shows a generic fuel consumption graph for a reciprocating engine. Normally the graph has two curves for the predicted fuel consumption. One curve is for 100% torque with varying speed, and the other curve is for 100% speed with varying torque. 3rd degree polynomial equations were written to represent each of these curves. These equations are stored in a program, one for each type of reciprocating engine.

From Fig. #1, at 100% speed and 100% torque, the fuel consumption is 7000 btu/bhp-hr.[2] At 100% speed and 70% torque, the fuel consumption is 7600 btu/bhp-hr.

At reduced speeds and torques, the predicted fuel consumption is actually between the 100 % speed/varying torque curve and the 100% torque/varying speed curve. Equations were written to interpolate data between the two curves to predict fuel consumption at reduced speeds and reduced torques.

[2]The manufacturer's fuel consumption curves are stated in heat rate (but/bhp-hr), not fuel consumption (MCF/D). By definition, the two terms are different. For reading simplicity, heat rate is referred to as fuel consumption in this paper.

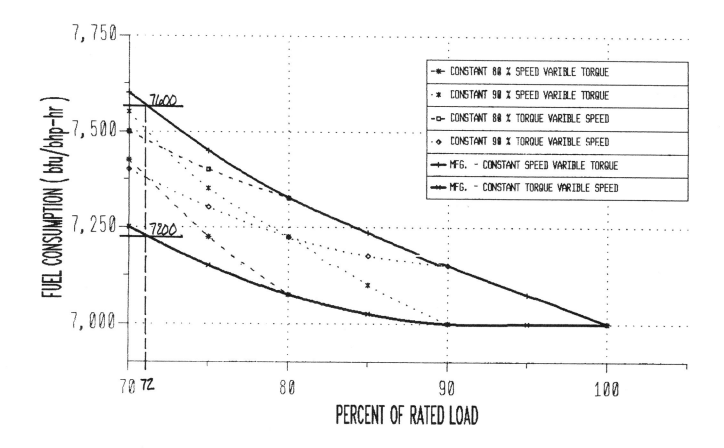

FIG. #2 - PREDICTED FUEL CONSUMPTION FOR OFF-RATED CONDITIONS

Using Fig. #2, for an engine operating at 90% torque and 80% speed, the predicted fuel consumption is 7,333 btu/bhp-hr. The percent torque and percent speed are multiplied to obtain percent load: 90% * 80% = 72%. The predicted fuel consumption from the constant torque curve (7200 btu/bhp-hr) and the constant speed (7600 btu/bhp-hr) are obtained by running the stored polynomial equations at 72%. The constant torque number and constant speed number from the polynomial equations are multiplied by ratios calculated from a weighted average of percent torque to percent speed. These numbers are then added together to determine the final predicted fuel consumption rate.

Weighted Average Calculation

(100% - % speed) + (100% - % torque) = Z

X = (100% - % torque) / Z

Y = (100% - % speed) /Z

Constant * X + constant * Y = Predicted Fuel
  speed   torque   Consumption

(100 - 80) + (100 - 90) = 30

X = (100-90)/30 = .3333

Y = (100-80)/30 = .6667

7600 * .3333 + 7200 * .6667 = 7,333 btu/bhp-hr

149

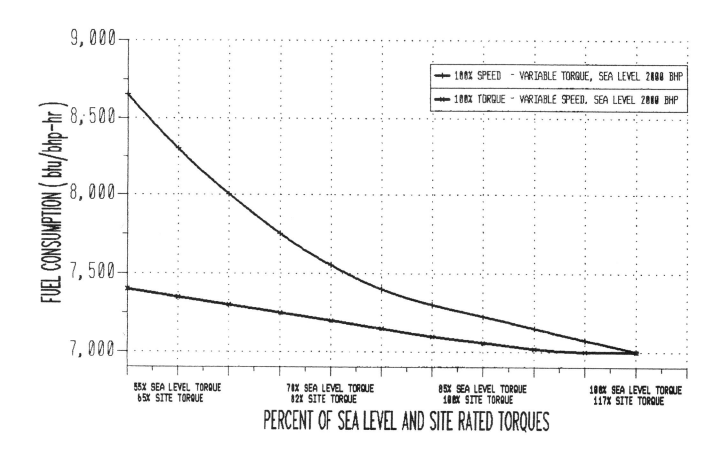

FIG. #3 - ADJUSTMENT FOR ELEVATION

ALTITUDE

Manufacturers' fuel consumption curves are for the sea level rating of the engine. Most of Northwest Pipeline's engines are significantly above sea level, and engine horsepower is derated to compensate for the decreased air density. Therefore, the manufacturers' curves must be adjusted to accurately predict the fuel consumption at altitude.

These curves are adjusted by multiplying the ratio of the site rated horsepower to the sea level rated horsepower. As an example, a reciprocating engine has a sea level rating of 2000 bhp, and is operating at an altitude of 6000 feet above sea level. The engine would be derated to approximately 1700 bhp. Therefore, operating at 100% site rated load of 1700 bhp and rated speed, the engine is operating at 85% (1700 bhp/2000 bhp) of the sea level rated torque. The predicted fuel consumption curve would need to be adjusted by 15% to account for the altitude adjustment. (Fig. #3) Operating at 100% site rated load of 1700 bhp and 100% speed, the predicted fuel consumption would be 7100 btu/bhp-hr.

ACTUAL FUEL CONSUMPTION MEASUREMENT

An individual meter run for each engine measures the fuel flow. The American Gas Association Report #3 (AGA 3) calculation is used to total the fuel flow to the engine. This calculation uses several factors to calculate fuel flow. Differential pressure, static pressure, and specific gravity factors are measured variables. The remainder of the factors are assumed to be constant. A SCADA system calculates the fuel consumption of each unit and logs this in a data base. The fuel consumption is stored in the data base in MCF/D. For use in this fuel report, this MCF/D is converted to btu/bhp-hr.

btu/bhp-hr = ((MCF/D * HHV) * / bhp) * 37.5

37.5 = (1000 cu.ft./M *.9 LHV/HHV) / 24 hr/day

HHV = higher heating value of the gas
LHV = lower heating value of the gas

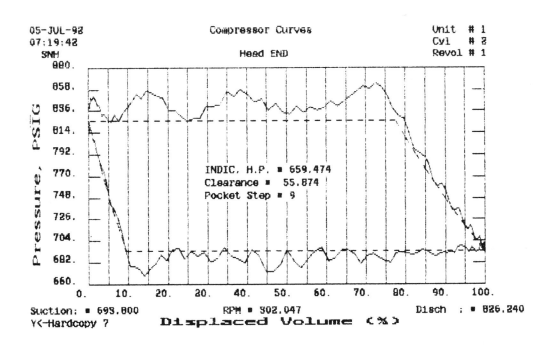

FIG. #4 - PRESSURE/VOLUME CURVE

The btu IIIIV value of the gas and the engine load is measured and stored hourly in the data base. This value is measured by an on-line chromatograph. The engine load is measured by Kistler pressure transducers. These are piezoelectric transducers that measure pressure change. Transducers are located in the head end and crank end of each compression cylinder.

A pressure/volume curve is generated during the compression cycle. (Fig. #4) The area inside the curve represents the work or load. The load calculated from all of the pressure/volume curves from each head end and crank end are added together to sum the total compression load on the engine.

Parasitic horse power is estimated and added to this load. This includes oil pumps, hydraulic fluid pumps, etc.

The percent engine torque is calculated by taking this total load, dividing by the site level rating of the engine, and multiplying by a ratio of the rated speed over operating speed.

$$\% \text{ torque} = \frac{\text{calculated load}}{\text{rated load}} * \frac{\text{rated speed}}{\text{operating speed}}$$

151

FUEL CONSUMPTION REPORT

The fuel consumption report is comprised of two sections. One section consists of engine operating data extracted from the data base. Another section is a management summary report.

Table #1 shows the section of the report which gives engine operating data. A query is run to select this information. Constraints are used in the program to select credible operating data. By using these limits, erroneous data is minimized and data is gathered within the predictable limits of the manufacturers' fuel consumption curves.

Table #2 shows another section of the report which condenses the data to give management the pertinent information required to interpret engine fuel consumption. The report lists average engine torque, average total bhp, number of hourly points used for the averaging, the average thermal efficiency, the average fuel consumption rate, and the average deviation of actual fuel consumption versus predicted fuel consumption.

This summary report flags an engine when the actual fuel consumption is 10% over the predicted fuel consumption. Another flag is displayed for each 10% increment, up to 50%. The summary report will also flag thermal efficiencies below 25% and above 40%, and fuel consumption rates below 6,300 btu/bhp-hr and above 12,000 btu/bhp-hr. Flagging abnormal thermal efficiencies and fuel consumption rates will indicate either erroneous data or an inefficient engine.

BENEFITS OF THE REPORT

The fuel consumption report has flagged several engines that were operating 10% over the predicted fuel consumption. Deficient fuel valves, fowled spark plugs, bad transmitters, incorrect software factors, wrong top dead center offsets, and numerous other mechanical problems were pinpointed and corrected based on the fuel report. Below are two actual examples of problems flagged by the fuel report.

In one instance, a 4000 bhp engine was flagged for operating 30% over the predicted fuel consumption. The engine was found to be operating out of balance. The engine was balanced and the actual fuel consumption dropped to 5% over the predicted fuel consumption. A fuel saving of $8,000 per month was realized.

In another case, one station had four 2000 bhp engines that were consistently operating 10% to 15% over the predicted fuel consumption. The engine was in excellent operating condition, but was difficult to balance properly. After months of checking for the problem, an engine analyzer was utilized to help solve the problem. The engine was actually developing 10% more horsepower than the computer calculated. This miscalculation lead to engine overload and detonation. A software factor in calculating engine load was found to be incorrect. This was corrected, the detonation ceased, and actual fuel consumption dropped to 2% over the predicted fuel consumption.

The actual and predicted fuel consumptions are now displayed on the local compressor station's CRT screen. This allows the station operators to compare the actual and predicted fuel consumption daily and repair any problems immediately if required.

The fuel report flags inefficient engines caused by improper air/fuel or ignition control, resulting in high NOX or CO_2 pollutants. By resolving these inefficiencies, the environment benefits by less air emission pollutants.

CONCLUSIONS

With the modern technology of gathering, storing, and manipulating engine operating data, programs can be written to evaluate enormous amounts of engine operating data in a matter of seconds. This data output can then be formatted in a concise report form that is a useful tool in evaluating operating efficiencies, such as the fuel report. Management now has information available to check the thermal efficiency and the condition of NWP's engines.

The results also indicate if there is a computer software or hardware problem, or if any of the hardware devices mounted on the engine that measure operating data is failing. These hardware devices are the same devices that control the engine loading and the engine air/fuel ratio control. Therefore, the fuel report adds extra insurance that the loading and air/fuel controls are working properly.

With the continuous monitoring of the specific fuel consumption, inefficient engines are flagged and corrected immediately, saving fuel cost and possibly costly engine mechanical failures. These saving are passed on to our customers, lowering our cost of service and increasing Northwest Pipeline's competitive edge.

RECIPROCATING ENGINE FUEL EFFICIENCY REPORT

LOCATION	CALENDAR DATE	TIME	UNIT	RPM	SCADA % LOAD	FUEL FLOW (MCFD)	FUEL HHV BTU	FUEL % LOAD	COMPR BHP LOAD	TOTAL BHP LOAD	ACTUAL SPECIFIC FUEL USE (BTU/HP-HR)	PREDICTED SPECIFIC FUEL USE (BTU/HP-HR)	ENGINE THERMAL EFF %	DEVIATION OF FUEL (ACTUAL / PREDICTED) %	FUEL USE DEVIATION OVER 10%
MOA	05/01/92	0100	04	301	101.3	324	1065	88.4	1748	1773	7298	7143	34.9	2.2	
MOA	05/02/92	0100	04	298	100.0	324	1055	87.3	1709	1734	7394	7163	34.4	3.2	
MOA	05/03/92	0100	04	301	99.6	318	1058	86.9	1719	1744	7235	7165	35.2	1.0	
MOA	05/04/92	0100	04	301	102.0	323	1053	89.0	1760	1785	7144	7135	35.6	0.1	
MOA	05/05/92	0100	04	301	101.9	315	1069	88.9	1759	1784	7080	7136	35.9	0.8	
MOA	05/06/92	0100	04	300	103.8	325	1060	90.6	1785	1810	7136	7116	35.7	0.3	
MOA	05/07/92	0100	04	233	100.1	245	1054	87.3	1337	1362	7109	7509	35.8	-5.3	
MOA	05/08/92	1500	04	300	106.5	333	1049	92.9	1832	1857	7055	7091	36.1	0.5	
MOA	05/09/92	0100	04	262	90.5	308	1054	79.0	1359	1384	8793	7612	28.9	15.5	$
MOA	05/10/92	0100	04	231	89.5	275	1045	78.1	1185	1210	8904	8024	28.6	11.0	$
MOA	05/11/92	0100	04	301	104.7	342	1049	91.4	1807	1832	7344	7108	34.7	3.3	
MOA	05/12/92	0100	04	281	101.5	309	1047	88.6	1635	1660	7307	7175	34.8	1.9	
MOA	05/13/92	0100	04	268	101.4	287	1051	88.5	1558	1583	7145	7225	35.6	1.1	
MOA	05/14/92	0100	04	248	100.2	270	1050	87.4	1425	1450	7333	7377	34.7	-0.6	
MOA	05/15/92	0100	04	238	99.5	267	1047	86.8	1358	1383	7582	7484	33.6	1.3	
MOA	05/16/92	0100	04	298	98.7	323	1047	86.1	1686	1711	7411	7182	34.3	3.2	
MOA	05/17/92	0100	04	301	98.7	330	1043	86.1	1703	1728	7468	7178	34.1	4.0	
MOA	05/18/92	0100	04	248	101.3	275	1039	88.4	1440	1465	7312	7344	34.8	0.4	
MOA	05/19/92	0100	04	241	100.8	270	1040	87.9	1393	1418	7427	7413	34.3	0.2	
MOA	05/20/92	0100	04	232	99.9	249	1045	87.2	1329	1354	7208	7526	35.3	-4.2	
MOA	05/21/92	0100	04	272	101.8	301	1052	88.8	1588	1613	7364	7200	34.6	2.3	
MOA	05/22/92	0100	04	234	100.9	263	1048	88.0	1354	1379	7497	7470	33.9	0.4	
MOA	05/23/92	0100	04	233	102.2	257	1047	89.2	1365	1390	7258	7434	35.1	2.4	
MOA	05/24/92	0100	04	274	100.9	295	1052	88.0	1585	1610	7228	7210	35.2	0.2	
MOA	05/25/92	0100	04	257	102.7	289	1047	89.6	1513	1538	7376	7250	34.5	1.7	
MOA	05/26/92	0100	04	233	100.9	262	1051	88.0	1348	1373	7521	7480	33.8	0.6	
MOA	05/27/92	0100	04	233	100.6	259	1049	87.8	1344	1369	7443	7232	34.2	0.6	
MOA	05/28/92	0100	04	232	102.7	255	1052	89.6	1366	1391	7232	7426	35.2	2.6	
MOA	05/29/92	0100	04	231	99.4	250	1049	86.7	1316	1341	7331	7555	34.7	3.0	
MOA	05/30/92	0100	04	237	101.9	252	1054	88.9	1385	1410	7066	7410	36.0	4.6	
MOA	05/31/92	0100	04	232	101.3	250	1052	88.4	1347	1372	7186	7474	35.4	3.9	
MTH	05/01/92	0100	01	328	91.6	397	1027	87.5	1716	1741	8781	7584	29.0	15.8	$
MTH	05/02/92	0800	01	328	99.3	429	1026	94.8	1860	1885	8754	7444	29.1	17.6	$
MTH	05/03/92	0100	01	298	97.9	382	1033	93.5	1666	1691	8748	7513	29.1	16.4	$
MTH	05/04/92	0100	01	298	77.9	302	1032	74.4	1326	1351	8651	8063	29.4	7.3	
MTH	05/05/92	0100	01	298	74.5	288	1034	71.1	1268	1293	8636	8187	29.5	5.5	
MTH	05/06/92	0100	01	298	86.5	344	1030	82.6	1472	1497	8873	7788	28.7	13.9	$
MTH	05/07/92	0100	01	298	91.9	373	1028	87.8	1564	1589	9047	7644	28.1	18.4	$
MTH	05/08/92	0100	01	328	97.0	421	1031	92.6	1817	1842	8835	7479	28.8	18.1	$
MTH	05/09/92	0100	01	299	73.7	305	1029	70.4	1259	1284	9168	8215	27.8	11.6	$
MTH	05/10/92	0100	01	301	83.5	328	1033	79.7	1436	1461	8699	7870	29.3	10.5	$
MTH	05/14/92	1000	01	299	83.3	325	1037	79.6	1423	1448	8730	7881	29.2	10.8	$
MTH	05/15/92	0100	01	298	84.3	331	1037	80.5	1435	1460	8816	7853	28.9	12.3	$
MTH	05/16/92	0100	01	299	82.4	324	1037	78.7	1407	1432	8797	7910	28.9	11.2	$
MTH	05/17/92	0100	01	328	93.9	424	1037	89.7	1759	1784	9241	7535	27.5	22.6	$$
MTH	05/18/92	0100	01	327	87.6	390	1037	83.7	1636	1661	9129	7684	27.9	18.8	$
MTH	05/19/92	0100	01	300	82.0	311	1032	78.3	1405	1430	8416	7920	30.2	6.3	
MTH	05/20/92	0100	01	328	90.4	395	1035	86.3	1694	1719	8920	7612	28.5	17.2	$
MTH	05/08/92	0800	02	286	79.9	325	1040	76.3	1305	1330	9528	8027	26.7	18.7	$
MTH	05/09/92	0100	02	281	83.4	333	1029	79.6	1339	1364	9423	7928	27.0	18.9	$
MTH	05/10/92	1200	02	283	94.4	370	1037	90.2	1526	1551	9277	7619	27.4	21.8	$$
MTH	05/20/92	2300	02	286	81.5	331	1034	77.8	1331	1356	9462	7975	26.9	18.7	$
MTH	05/21/92	0100	02	279	82.3	323	1035	78.6	1312	1337	9379	7968	27.1	17.7	$
M1H	05/22/92	0100	02	282	78.8	307	1033	75.3	1269	1294	9188	8076	27.7	13.8	$

TABLE #1 - DETAILED FUEL EFFICIENCY REPORT

LOCATION	CALENDAR DATE	UNIT	HOURLY DATA PTS USED FOR AVERAGE	AVG SCADA % TORQUE	AMBIENT SITE RATED BHP	AVG BHP LOAD	AVG DEV OF FUEL (ACTUAL / PREDICTED) %	AVG THERMAL EFF	AVG HEAT RATE	FUEL USE DEVIATION OVER 10%
BAK	05/92	1	6	83	1450	1229	-1.0	31.5	8143	
BAK	05/92	2	6	93	1450	1377	-2.5	34.1	7474	
BAK	05/92	3	7	85	1450	1248	0.3	31.4	8149	
BAK	05/92	4	7	95	1800	1741	-3.9	38.4	6631	
BPY	05/92	1	14	89	836	783	19.3	25.3	10077	$
BPY	05/92	2	14	83	836	732	23.7	24.0**	10637	$$
BPY	05/92	3	13	80	678	582	4.2	26.8	9500	
BPY	05/92	5	13	88	507	473	5.9	27.0	9416	
BPY	05/92	6	14	86	507	464	-2.4	29.2	8722	
BPY	05/92	7	13	83	507	452	2.0	27.7	9189	
BPY	05/92	8	13	74	507	401	-4.2	28.4	8958	
BPY	05/92	9	14	89	507	480	42.1**	20.2**	12599.**	$$$$
BPY	05/92	11	14	110	4000	4462	2.1	35.6	7164	
BUR	05/92	1	5	84	1825	1546	1.7	33.8	7540	
BUR	05/92	2	11	94	1825	1655	-2.8	36.2	7031	
BUR	05/92	3	15	91	1825	1596	-2.8	35.6	7149	
CAL	05/92	1	28	95	1450	1398	5.7	31.7	8026	

NOTE: 1) THE FUEL USE REPORT IS BASED ON PICKING THE SECOND LEGITIMATE HOURLY DATA POINT EACH DAY.

    % LOAD BETWEEN 70% & 130% OF RATED HORSEPOWER
    BTU'S BETWEEN 950 & 1150
    RPM'S > 200
    FUEL FLOW > 50 MCF/D

2) % LOAD FOR FUEL MODIFIED TO MANUFACTURER'S SEA LEVEL FUEL CONSUMPTION CURVES.

3) ACTUAL FUEL RATE IS REDUCED BY PARASITIC HORSEPOWER

4) PREDICTED FUEL IS CALCULATED USING BOTH CONSTANT SPEED AND CONSTANT TORQUE CURVES

5) THE COOPER UNITS AT GREEN RIVER AND KEMMERER USE GMWA & GMWC FUEL CONSUMPTION

6) THE COOPER GMWC & GMWA ARE ASSUMED TO HAVE CONSTANT TORQUE CURVES WITH SIMILAR SHAPE TO THE GMVII

** THERMAL EFF < 25% OR > 40% AND HEAT RATE < 6500 OR > 12,000 NOT USED IN OVER-ALL AVERAGES.
** AVERAGE DEVIATION OF FUEL < 10% OR > +25% ALSO NOT USED IN OVER-ALL AVERAGES.

TABLE #2 - REPORT SUMMARY

TWO-STROKE ENGINE WITH DISINTEGRATED CYCLE

Radislav Pavletič and Dejan Leskovšek
Department of Mechanical Engineering
University of Ljubljana
Ljubljana, Slovenia

ABSTRACT

Compression phase of the I.C. engine cycle was transmitted from the cylinder to its external intake system. Combustion chamber is thus filled with the pre-compressed mixture. Control of the gas flowing from the chamber to the cylinder space is controlled by a special transflow valve or by the engine piston alternatively. Volume of the tubular combustion chamber can be altered even during the operation of the engine and this is performed by a special plunger, which displacement is linked with the main control loop of the engine. Alteration of the combustion chamber volume introduces overexpansion at partial engine load. Cyclic combustion variations of the tubular shaped combustion chamber, as well as its tendency to knock was observed on a model engine and have given promising results.

In spite of great efforts that have been made during the last ten years in the development of conventional two-stroke engines, the required results have not been obtained. Working concept was mainly based on the simplicity of design. Sophisticated but more expensive technical solutions that could better satisfy the demands of the modern vehicle engines, i.e.: fuel economy, environment protection etc. unfortunately exceed the advantages of the simplicity of its design.

Sometimes it is reasonable to sacrifice some cost-effectiveness to satisfy better performances of the engine. However, extra costs should be fully justified through all the development phases: from the basic principles, modeling and necessary performance tests to the construction of the prototype.

DISINTEGRATION 0F THE CYCLE

Combustion of pre-compressed working medium is one of the main conditions for an effective performance of the thermodynamic cycle within the I.C.engine. Condition of the working mixture at the end of the compression phase greatly affects the quality of the working process; it is determined by the state of the mixture at the beginning of the compression - i.e. by the effectiveness of the filling process, the heat exchange during the compression, leakage of the mixture, etc. Parameters of the compressed working medium are therefore affected by previous events and can not be selected arbitrarily. They are also limited by possible disturbances that could take place during the process of combustion. The way, proper in which state of the gas before the ignition is obtained, is of no consequence. It is therefore possible to perform the compression separately and fill the combustion space with pre-compressed medium. Working parameters such as pressure and temperature can thus be better controlled. Engine of that type can be called engine with disintegrated cycle.

CONCEPT OF THE ENGINE

The design of the engine with disintegrated cycle requires a new intake-compression system together with some modifications of the engine construction. Charging process in the combustion chamber must be concluded before the ignition is indicated. In the case of a conventional crank mechanism the combustion chamber should be separated from the piston stroke volume during the intake process by a special controlled valve. Such a solution is schematically shown in Fig. 1 together with valve timing diagram. Besides the timing of the combustion chamber-cylinder volume flow control valve (transflow valve), diagrams for the inlet of the pre-compressed mixture in the combustion chamber and the conventional engine like exhaust valve timing diagram are shown in Fig.1 too. Transflow valve of a disintegrated cycle engine is a novelty that, due to its required operational reliability at higher thermal loads, brings some additional design complications to the timing mechanism.

Omission of the transflow valve is possible; its function could be replaced by the motion of the piston. Piston should in this case be delayed in the top dead center for some extra time. Such mechanisms to enable this particular piston motion are already known; inverse crank mechanism with short connecting rod could be a good solution and is schematically shown, together with timing diagram for the engine in Fig 2. Separated combustion chamber controlled by a valve or by the engine piston itself can be cylindrical or even tubular shaped. Alteration of the chamber volume is obtained by a moving control plunger in it; the pre-compressed charge flow and thus the engine load are therefore influenced by the position of that plunger. The plunger action can be integrated into the main engine control system; combustion chamber volume can be altered even during the operation of the engine. It determines mass flow rate at known parameters of the pre-compressed charge and consequently the load of the engine. Smaller combustion chamber volume at a partial engine load means also prolonged expansion and therefore better thermodynamic efficiency.

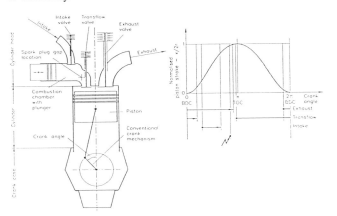

Figure 1. Engine and timing diagram; transflow valve controls burning gas flow

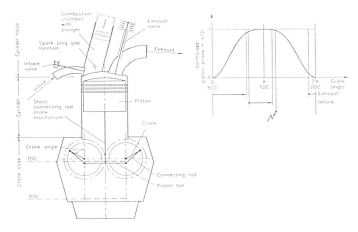

Figure 2. Engine and timing diagram; engine piston controls the burning gas flow

An engine with fully developed inletcompression assembly is schematically shown in Fig. 3. It consists of a turbocharger (1) as the first stage of the compressor chain, (2) a mechanically driven compressor of volumetric type or additional compressor stages for higher pressure levels, (3) intercoolers for the preferably isothermal compression, (4) a heat exchanger for regenerative pre-heating of the inlet medium with exhaust gases heat and (5) variable driving mechanism for the volumetric compressors. For a modest configuration of the inlet-compressor assembly a direct driven volumetric compressor with a throttle valve is indispensable. Introduction of other components mentioned above increases the efficiency and complexity of the integral engine. As a compromise, a slightly more sophisticated solu-

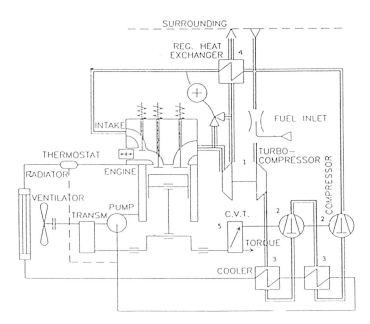

Figure 3. Maximum configuretion of inlet-compression system of the engine with disintegrated cycle

tion together with its main control loops is shown in Fig. 4. Control loops are determined by: the change of the combustion chamber capacity, the change of the charge temperature, by the alteration of the flow of the exhaust gases through the heat exchanger, and by the alteration of the pressure level determined by the speed of the compressors. The above mentioned control parameters influence the load as well as the ecologic picture of the engine.

From the standpoint of the combustion process quality, there is an evident advantage of the disintegrated-cycle over the conventional two- stroke cycle engine. The construction of the new type engine is more sophisticated but made of components, that

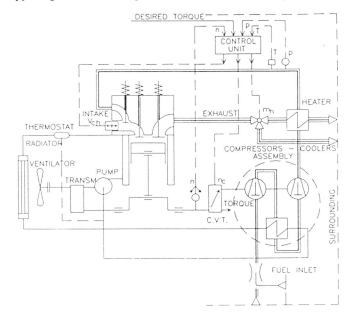

Figure 4. Reduced configuration of the inlet-compression system of the engine with disintegrated cycle

156

have already been used in the engine technology. Some more questions arise concerning the design of the volumetric compressors for higher pressure levels. The experiences we have about mechanically driven low pressure compressors for supercharged engines cannot be simply used for higher pressure stages; new research efforts must be made to develop compressor units or cascades for the approximately 2 MPa pressure level with integrated and nearly isothermal cooling of the medium with the total efficiency of near 80%.

ENGINE PERFORMANCES-COMPUTED

Performances of the engine with disintegrated cycle according the Fig.4 were calculated on the basis of a zero-dimensional analytical model and presented by its efficiency and cycle working capability in Fig. 5 (Pavletič (1990)). Performances of the indicated process are presented in the upper part of Fig.5; work required for the external compression and for the transport of the working medium into the combustion chamber has not been subtracted. Particular curves correspond to diverse ways of load changes. The upper bold envelope declares the total field of indicated processes; the right hand branch corresponds to the engine loads caused by the alteration of the combustion chamber volume at the prescribed parameters of the intake mixture. Central section of the envelope takes into account engine loads at the reduced combustion chamber volume and different temperatures but at the same pressure of the fresh charge, whereas in the left side branch influences of different levels of the charge pressure are considered. Characteristic working field is limited by two dotted curves with a common starting point-the point of the maximum load. The upper curve defines the change of the engine load by the alteration of the combustion chamber volume; inlet pressure of the working medium has its maximum value, while its temperature is assumed to be at the minimum. The lower limiting curve describes the situation where the engine load corresponds to the inlet pressure alterations; combustion chamber volume is at its maximum value and the inlet temperature of the charge at its minimum. The advantage of the load control by means of the combustion chamber volume altera- tion is evident.

Effective parameters of the engine (subtracted extra work required for the external compression, medium transport and engine friction loses) were calculated and are presented in the lower part of fig.5. Diverse values of the external compression efficiencies and diverse ways of the engine load alteration (according to the envelope in the upper part of the diagram) were also taken into account. Reduction of the total compressor efficiency as well as the deviation from the pure isothermal compression substantially reduce effective parameters of the newly conceived engine. Compressor efficiencies of near 80% and almost perfect intercooling ensure compatitive performances of the new engine when compared with the modern conventional I.C.engines. One can see strong influence of the above mentioned two parameters on the engine effective performances in figure 6. The engine torque characteristic has its maximum value at the lowest engine speeds. Minimum specific fuel consumption is found at lower engine speeds and loads. This position i very attractive for a vehicle driving engine.

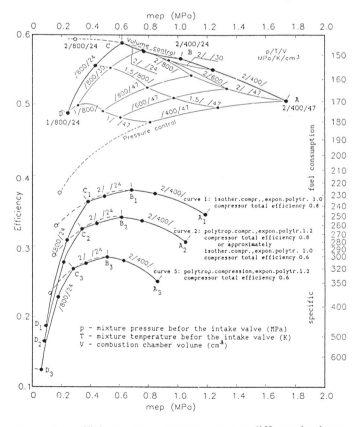

Figure 5.   Efficiency-mean pressure map at different load control (calculated)

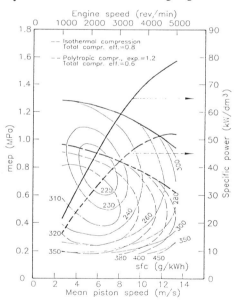

Figure 6.   Engine performance map (calculated)

MODELING IN RESEARCH WORK

The main features of the presented engine concept require additional analytical, numerical simulations and experimental work with suitable physical models. Phenomena of combustion in a tubular shaped combustion chamber were observed on an experimental model-engine. Three types of models simulating combustion chamber with transflow valve (Fig.1) or piston controlled combustion chamber (Fig.2) were studied on a basic two-stroke,80 ccm swept volume, conventional petrol engine with modified cylinder heads according to Fig. 7. The rasults obtained

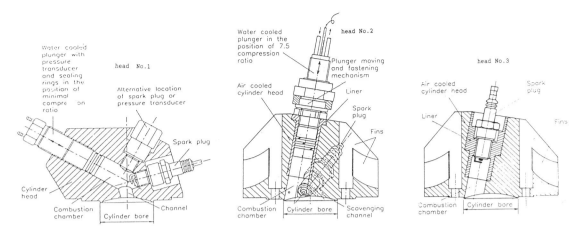

Figure 7. Different cylinder heads with tubular combustion chambres

with the cylinder head No.1 were presented by Pavletič (1991), while the rasults obtained with the cylinder head No.2 were presented by Leskovšek (1991), which together with the results obtained with the cylinder head No.3 will be briefly explained here. The differences between three heads regarding the combustion chamber connecting nozzle shape and spark plug positions can be recognised from Fig. 7.

The main aim of the study was to compare cyclic variations of the combustion process and sensitivity to knock appearances for the conventional opened hemispherical chamber with the compression ratio 7.5 and two tubular combustion chamber heads with the spark plug positioned, first, right in the nozzle connecting the combustion chamber and cylinder space - head No.2 and second in the bottom of the chamber opposite to the nozzle - head No.3. For the tubular shaped combustion chamber heads the compression ratios were altered too. The octane number of fuel composed by the referential fuels (iso-octane and n-heptane) varied from 0 to 80.

The cyclic variation of particular combustion processes is represented by the standard deviation of the peak cylinder pressure versus octane number and is shown in Fig. 8 for the production engine cylinder head and for the new designed cylinder heads designated as head No.2 and head No.3 (Fig.7). The standard deviation is proportionally lower for tubular shaped combustion chamber head with the spark plug positioned in the chamber nozzle (head No.2), but higher for the tubular shaped combustion

chamber head with the spark plug in the bottom of the chamber (head No.3) for the hole range of the fuel octane number and for both engine speed considerate. Scattering of the cyclic variations is connected with the duration of the combustion process. Duration of the combustion process is presented in the Fig. 9 for two engine speeds and for all three types of combustion chambers; additional results for the tubular combustion chamber heads are shown for two more compression ratios. Ignition time was selected according to the best working capability of the cycle (MBT); in Fig.9 one can select the beginning of the ignition and the time in which the value of the peak pressure is reached. This time difference corresponds to the period of intensive combustion and is again shorter for the tubular combustion chamber with the spark plug positioned in the chamber nozzle and longer for the spark plug positioned in the bottom of the chamber. Cyclic variations are less pronounced in the cases with more intensive first phase of combustion. Disposition of the process to knock is shown in Fig. 10. As the measure of its intensity, mean statistical value of the pressure increase rate was considered as a suitable parameter. Two diagrams in the upper part of fig.10 represent comparative disposition to the knock for the conventional and the tubular combustion chamber with the nozzle positioned spark plug (head No.2) at the same compression ratio. Lower diagrams show suitable results for the tubular shaped combustion chamber and higher compression ratios. We can conclude that the velocity of the pressure rise is less intense for the tubular combustion chamber throughout a range of diverse fuel octane numbers. Lower diagrams show that the intensity of the pressure

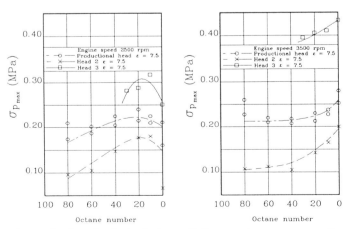

Figure 8. Standard deviation of the peak pressure as the measure for cyclic variation of combustion

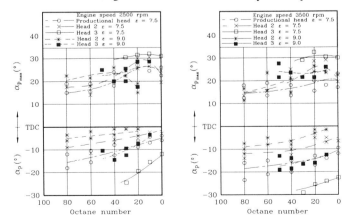

Figure 9. "Combustion rate" - angles of ignition and cycle peak pressures

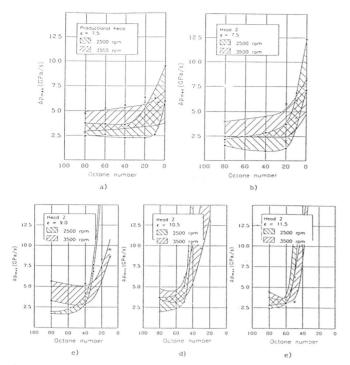

Figure 10. Knock appearance

rise moves towards higher octane number when higher compression ratios are considered. At the engine speed of 2500 r.p.m. and the maximum value of the compression ratio the appearance of the knock disturbances takes place only at the octane number 60 approximately. But, surprisingly, we didn't find any significant differences regarding the knock inclination between the tubular shaped combustion chambers at different spark plug positions.

CONCLUSION

There is a certain advantage of the two stroke model engine with disintegrated cycle when compared to the conventional one. Preliminary modeling and the accompanying tests have proved that the tubular shaped combustion chamber shows better resistibil-

ity to knock in a great deal also independently from the spark plug position. On the other side, combustion stability expressed with less intensive cyclic peak pressure variation with tubular shaped combustion chamber is significantly improved if the spark plug is positioned in the combustion chamber nozzle. In the opposite case, if the spark plug is positioned in the bottom of chamber opposite to its nozzle, the cyclic variation is worse.

Problems generally arise in connection with the construction of components that haven't yet been satisfactorily solved. Besides, the new concept of the combustion chamber introduces a new quality parameter: simultaneous alteration of the combustion chamber volume during the engine operation that makes possible overexpansion and therefore better efficiencies at lower loads. One must remember that one of the main disadvantages of a conventional petrol engine is the inefficient load control by throttling. Good resistibility to knock of the tubular combustion chamber allows extremely high values of engine load even at the lowest speeds. Torque characteristic can therefore be improved and the engine elasticity as well. Compromise between the attractive and very simple conventional two-stroke petrol engine and the new concept engine with its good combustion stability and better efficiency is destroyed by the fact that, there are no yet suitable technical solutions available for the inlet system with integrated pre-compression.

ACKNOWLEDGEMENT

The support provided by the Research Council of Slovenia is gratefully acknowledged.

REFERENCES

Pavletič R., Sinyavskiy V.V., Bizjan F., Leskovšek D.(1990): "I.C.engine having performance parameters preferential for automotive operating conditions"; Proceedings of COMODIA 90,Kyoto, pp.185-191

Pavletič R.,Leskovšek D. (1991): "The tubular combustion chamber for I.C.engines"; Proceedings of AIEC,Buenos Aires,

Leskovšek D.(1991): Combustion chamber tubular shaped - its suitableness for S.I.engines"; MSc work,Faculty of Mech.Eng.Univ. Ljubljana, Slovenija

ICE-Vol.18, New Developments in Off-Highway Engines
ASME 1992

DYNAMIC AND STATIC FLOW ANALYSIS
OF A GASOLINE FUEL INJECTOR

J. L. Chen, Grant Chen, and Marv Wells
Electrical and Fuel Handling Division
Ford Motor Company
Ypsilanti, Michigan

ABSTRACT

The effect of the electrical pulse width and the fuel temperature on dynamic flow and static flow rate of a gasoline fuel injector has been numerically and experimentally investigated. In numerical analysis, the physical domain covers the region from upstream of the valve seat to the injector exit. The three-dimensional unsteady Navier-Stokes equations in a curvilinear coordinate system are solved. Due to the needle movement in fuel injection, the physical domain is considered as a function of time. In the experimental study, the test stand consisting of a hydraulic system and an electrical system was designed to meet the requirements of Society of Automotive Engineers. The pulse width of 0.97 - 7.5 ms and the temperature of 20 - 100 °C were used to study the pintle injector performance. Predicted dynamic flow and static flow rates show higher values at a temperature of 80 °C which are consistent with the test results.

INTRODUCTION

The gasoline fuel injection system injects the fuel to the intake port of the engine. Various pulse widths can be sent to the fuel injector according to the actual engine operating conditions as sensed by an electronic control-unit. As such, the fuel injection system can have more uniform spray distribution, more rapid engine response to changes in throttle positions, and more precise control of air/fuel ratio during cold-start and engine warm-up [Heywood, 1988]. To achieve these benefits, the fuel injector is required to provide a wide dynamic flow range and a minute variation in stroke-to-stroke fuel quantity. For a small engine, the low-end linearity between the dynamic flow and the pulse width are especially important for the fuel injector, because it has to accurately meter a small amount of fuel at a short pulse.

To deliver the liquid fuel, the solenoid coil of the injector, activated by a pulsed electronic signal, generates a magnetic field to lift the needle. This creates a passage between the needle and the valve-seat. The fuel is driven by the regulated pressure gradient between the fuel rail and the injector exit. The fuel injector is required to deliver fuel with minimal stroke-to-stroke variation and a spray cone with good atomization. These are essential for smooth low idle, low raw emissions and good driveability for the cold engines.

To improve the fuel injector performance, several companies have been developing new concepts and modifying the designs. Greiner et al. (1987) at Robert Bosch GmbH developed a multi-hole injector to reduce buildup of deposit and a bottom-feed injector to handle under-hood high temperature. Andrighetti, et al. (1987) at Lucas CAV

employed a low-mass flat armature and an open-orifice valve seat to improve stability of air/fuel ratio. Okamoto et al. (1992) at Hitachi Ltd. developed a two-stream injector by implementing an adapter near the orifice to split the fuel flow. Though these developments provide interesting progress in improving the injector performance, some basic issues are still not resolved. No information is yet available regarding how the fuel temperature affects the flowfield and pressure distribution inside the fuel injector and how much of the fuel is delivered during the opening cycle, the fully open stage, and the closing cycle. This is partly because it is very difficult to probe the flowfield and pressure in the injector (e.g., a typical pintle diameter is 0.46 mm).

In view of this, a computational study was carried out to analyze the temperature effects on fuel delivery and to determine the mass flow at different stages of fuel injection. In the meantime, the test data of dynamic and static flow rate were employed to validate the numerical modeling. The fuel temperatures ranging from 20 - 100 °C and the pulse widths of 0.97 - 7.5 ms are employed to investigate the pintle injector performance at two different operating conditions, i.e., the steady and transient states. At the steady state operation, the valve of the fuel injector is held fully opened and the delivered fuel rate is called the static flow rate. In the transient state operation, the needle movement is controlled by the electronic pulse signal and the fuel mass delivered during a pulse is called the dynamic flow.

TEST FACILITY

The test stand used to evaluate the gasoline fuel injector consists of a hydraulic system and an electrical system (see Fig. 1) which are designed to meet the requirements of the Society of Automotive Engineers (SAE, 1992). The hydraulic system supplies the test fluid to the injector inlet port at a stable flow rate and pressure, whereas the electrical system is to provide electrical pulses in controlling the valve movement of fuel injector(s).

The hydraulic system is capable of handling test fluids as n-heptane, indolene, and mineral spirits. During the test, the test fluid in the fluid reservoir is pumped out to the fluid line by a recirculating pump. The fluid pump is capable of delivering 20 gm/sec of fluid at a stable pressure of 700 kPa. The excessive test fluid is sent back to the fluid reservoir through a bypass fluid line under the control of a pressure relief valve. To maintain the fluid temperature variation within 1 °C, two heat exchangers are installed in the main fluid line and the bypass line, respectively. Downstream of the heat exchanger, there is a filter capable of removing solid particulates greater than 5

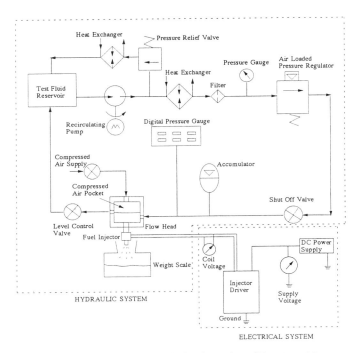

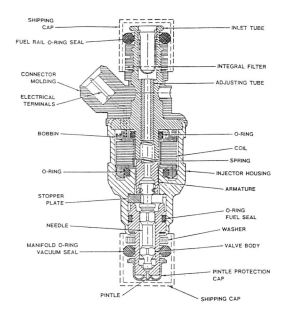

Fig. 2   Schematics of a Typical Pintle Gasoline Fuel Injector

Fig. 1   Hydraulic and Electrical Schematics of Test Stand for Gasoline Fuel Injectors

μm in order to avoid contamination in the fuel injector. The fluid pressure is reduced to the desired pressure by an air loaded pressure regulator which can achieve the accuracy of ± 0.1 kPa in the range of 40 - 500 kPa. The fuel pressure regulator is able to maintain the differential pressure across the injector at a value of 270 kPa and with a variation less than 0.5% through out the test. A shut-off valve is located between the regulator and the accumulator for emergency shut-down. To eliminate any effect of fluid head pressure on injector inlet pressure, the pressure gauge is installed in the same horizontal plane. In addition, an air dampened fluid head is employed to reduce the pressure pulsation caused by the pump. This is achieved by pre-charging air on the top of the injector mount to form an air pocket with a minimum volume of 30 CC at fuel system pressure. The spray generated by the injector is collected by a test fixture beneath the injector exit. A weight scale is employed to measure the mass of the fuel which drains into the fixture. In order to provide repeatable readings, the test fixture has to be totally wet down by a minimum of 10,000 pulses.

The electrical system consists of an injector driver, a DC power supply unit, and voltage meters. The injector driver is an electronic circuit which can supply voltage pulses to the fuel injector with an accuracy of ± 0.001 ms. Also, the time period elapsed from the beginning of the pulse to the next pulse is 10 ms. The DC power supply unit provides a voltage of 14 ± 0.05 VDC to the injector driver.

MATHEMATIC AND NUMERICAL MODELS

The pintle injector (see Fig. 2) is chosen in this study because of its simplicity in design near the injector exit and thus fewer parameters controlling the flowfield, which is more suitable for fundamental study. Also, better understanding of the pintle injector can provide more basic design guidelines for the advanced injectors. Prior to the fuel injection, the needle is pressed against the valve-seat by the helical spring to seal the flow passage. After the pulse is sent to the solenoid, it takes 0.9 ms of delay for the solenoid coil to generate enough force to overcome the spring force. The needle is therefore lifted to open the valve to start the opening cycle. The needle continues to lift and fully opens the valve at 1.45 ms which sets on the fully open region. When the pulse is over, the needle still holds its position for another

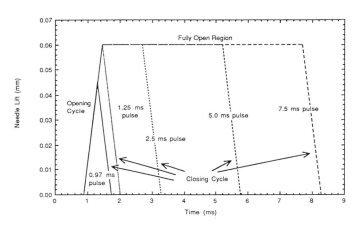

Fig. 3   Transient Needle Lift for Various Pulse Widths

0.2 ms because of the residual magnetic field in the coil. The valve then begins the closing cycle and takes additional 0.58 ms to fully close the valve which marks the end of fuel injection. The transient needle lifts at various pulse widths, ranging from 0.97 to 7.5 ms, are shown in Fig. 3.

The physical domain of the governing equations covers the region from upstream of the valve-seat to the injector exit (Fig. 4). This is because most of the pressure-drop is expected to occur in this region and the design of this region controls the accuracy of fuel-metering. The fluid dynamics in the injector is analyzed by solving the transient, three-dimensional Navier-Stokes equations. Since the flow temperature and density inside the injector maintain almost the same during the tests, the flow is considered as isothermal and incompressible in the computational analysis. The fuel density and kinematic viscosity at 20, 40, 60, 80, 100 °C are 7.34 x 10^{-4}, 7.24 x 10^{-4}, 7.1 x 10^{-4}, 6.97 x 10^{-4}, 6.83 x 10^{-4} gm/mm³ and 0.6, 0.5, 0.4, 0.32, 0.28 mm²/s, respectively. Both viscosity and density increase with temperature.

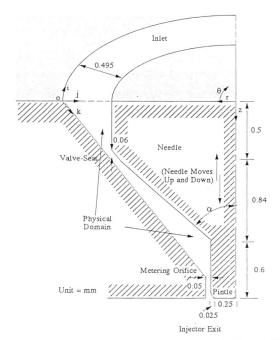

Fig. 4   Physical Domain of the Pintle Injector at the Needle Lift of 0.06 mm (Fully Open)

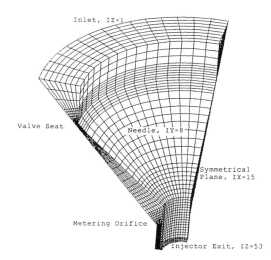

Fig. 5   Grid Distribution in the Computational Domain of the Fuel Injection

The Mean Flow Equations

The time-averaged equations for conservations of mass and momentum in a transient, three dimensional flow can be expressed in the tensor notation as

$$\frac{\partial U_i}{\partial X_i} = 0 \tag{1}$$

$$\frac{\partial(\rho U_i)}{\partial t} + \frac{\partial(\rho U_i U_j)}{\partial X_j} = -\frac{\partial P}{\partial X_i} + \frac{\partial}{\partial X_j}[\mu_l(\frac{\partial U_i}{\partial X_j} + \frac{\partial U_j}{\partial X_i}) - \rho\overline{u_i u_j}] \tag{2}$$

where the term $-\rho\overline{u_i u_j}$ is the Reynolds stress which stands for the contribution of turbulent motions to the mean stress.

The Turbulence Model

In the present work, the Reynolds stress is calculated by adopting the Boussineq turbulent eddy viscosity concept (Hinze, 1975) in which the Reynolds stress is expressed as

$$-\rho\overline{u_i u_j} = \mu_t(\frac{\partial U_i}{\partial X_j} + \frac{\partial U_j}{\partial X_i}) - \frac{2}{3}\delta_{ij}\rho k \tag{3}$$

where δ_{ij} is the Kronecker-Delta function and μ_t is the turbulent viscosity. The k-ε two-equation model (Launder and Spalding, 1974) is employed to calculate the turbulent viscosity, which the turbulent viscosity is expressed in terms of turbulent kinetic energy, k, and turbulent dissipation energy, ε, as follows

$$\mu_t = C_\mu \rho \varepsilon^2 / k \tag{4}$$

where $C_\mu = 0.09$. According to the two-equation turbulence model, the governing equations for k and ε are expressed as

$$\frac{\partial(\rho k)}{\partial t} + \frac{\partial(\rho k U_j)}{\partial X_j} = \frac{\partial}{\partial X_j}[(\mu_l + \mu_t/\sigma_k)\frac{\partial k}{\partial X_j}] - \rho\overline{u_i u_j}\frac{\partial U_i}{\partial X_j} - \rho\varepsilon \tag{5}$$

$$\frac{\partial(\rho\varepsilon)}{\partial t} + \frac{\partial(\rho\varepsilon U_j)}{\partial X_j} = \frac{\partial}{\partial X_j}[(\mu_l + \mu_t/\sigma_\varepsilon)\frac{\partial\varepsilon}{\partial X_j}] - C_1\frac{\varepsilon}{k}\rho\overline{u_i u_j}\frac{\partial U_i}{\partial X_j} - C_2\rho\frac{\varepsilon^2}{k} \tag{6}$$

where $C_1 = 1.44$ and $C_2 = 1.92$ are further constants in this model.

Also, $\delta_k = 1.0$ and $\delta_\varepsilon = 1.217$ are turbulent Prandtl numbers for k and ε, respectively.

Grid Distribution

Due to the symmetry in the θ direction, only a quarter of the injector is considered as the physical domain in the numerical analysis. The whole physical domain is then divided to 15 x 8 x 53 cells for the computational domain (see Fig. 5). In order to simulate the realistic physical domain, a curvilinear coordinate system, the Body Fitted Coordinates (BFC), is employed to connect 8 adjacent nodes to form a six-faced cell [Sugiura et al., 1990]. The three coordinates in the BFC are i, j, and k and the origin is located at the intersection of the inlet region, the valve-seat plane, and the symmetrical plane (see Fig. 4). The BFC grids can be viewed as the squeezed, stretched, bent, and twist orthogonal grids. In each grid, the conservations of mass and momentums are employed to form a set of algebraic equations. However, the direction of the velocity resolute in the BFC coordinates is not always normal to the corresponding surface of the grid. Thus, the velocity used in the mass-flux formula involves not only u, but also v and w. This new mass-flux velocity (say at the east surface of the grid), u_e^*, can be expressed in terms of the local grid geometry. i.e.,

$$u_e^* = F_u u_e + G_u v_n + H_u w_h \tag{7}$$

where F_u, G_u, and H_u are the functions of the local geometry of the grid in the i, j, and k directions, respectively, and u_e, v_n, and w_h are the velocities at the east, north, and high surfaces of the grid. Detailed expression of the local geometry functions, F_u, G_u, and H_u can be referred to [Rosten and Spalding, 1986]. The effective mass flux across the east surface is then expressed as

$$m_e = \rho_e A_e u_e^* \tag{8}$$

Similar treatment can be applied to the west, north, south, high, and low surfaces of the grid.

Initial and Boundary Conditions

In the event of fuel injection, the physical domain is a function of time in the opening and closing cycles. The grid sizes at these two stages are allowed to stretch and contract corresponding to the needle velocity, whereas the grid number is fixed. At the same time, the wall velocity is introduced along the needle body as a result of needle movement. The wall velocity in BFC is expressed in terms of the surface inclination angle and needle velocity in the cylindrical coordinate system. The initial and boundary conditions are expressed as follows.

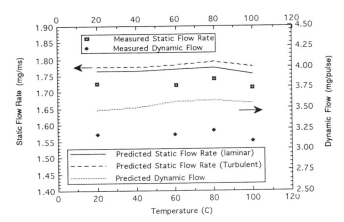

Fig. 6   Comparison of Predicted and Measured Dynamic Flow and Static Flow Rate at Various Temperatures with Pulse Width of 2.5 ms

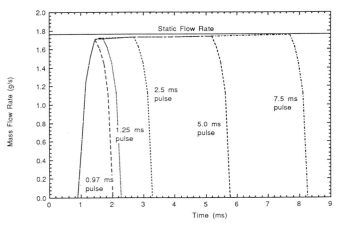

Fig. 7   Mass Flow Rate during Fuel Injection at Various Pulse Widths with Temperature of 20 °C

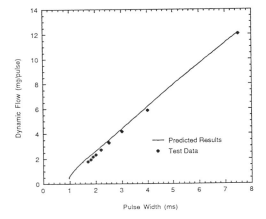

Fig. 8   Dynamic Flow at Various Pulse Widths and Temperature of 20 °C

Initial Conditions (at $t = 0$):

at upstream of the valve (i.e., $i = 0$ to $i = i_x$, $j = 0$ to j_y, and $k = 0$ to k_v)

$$u_i = v_j = w_k = 0, \quad p = p_f, \tag{9}$$

at downstream of the valve (i.e., $i = 0$ to $i = i_x$, $j = 0$ to j_y, and $k = k_v$ to k_z)

$$u_i = v_j = w_k = 0, \quad p = 0, \tag{10}$$

where k_v is the location of the valve seat in the k coordinate and p_f is the pressure at the fuel rail which is 270 KPa. The pressure value in present study is the referenced value with respect to the pressure at the injector exit.

Boundary Conditions (at $t > 0$):

at the inlet (i.e., $i = 0$ to i_x, $j = 0$ to j_y, and $k = 0$),

$$u_i = v_j = 0, \quad p = p_f, \tag{11}$$

Note that the value of the inlet velocity w_j in this study is not given and is part of the solution.

at the outlet (i.e., $i = 0$ to i_x, $j = 0$ to j_y, and $k = k_z$),

$$u_i = v_j = 0, \quad p = 0, \tag{12}$$

at the symmetrical planes (i.e., $i = 0$ and i_x, $j = 0$ to j_y, and $k = 0$ to k_z),

$$\partial u_i / \partial i = \partial v_j / \partial i = \partial w_k / \partial i \,, \tag{13}$$

at the valve-seat wall (i.e., $i = 0$ to i_x, $j = 0$, and $k = 0$ to k_z),

$$u_i = v_j = w_k = 0, \tag{14}$$

at the needle body wall (i.e., $i = 0$ to i_x, $j = j_y$, and $k = 0$ to k_z),

$$u_i = v_j = 0, \quad w_k = -w_d \cos(\alpha), \tag{15}$$

where w_d is the needle lift velocity and α is the inclination angle of the needle body surface in cylindrical coordinates.

The SIMPLEST algorithm and the staggered grid pattern are applied to integrate the finite-difference equations to obtain the flowfield and pressure distribution. The solvers of the PHOENICS computer code are applied to facilitate the numerical solution [Spalding, 1989].

RESULTS AND DISCUSSION

Two different operating conditions, i.e., the valve is held fully opened and the valve is controlled by a pulse ranging from 0.97 to 7.5 ms, with temperature of 20 - 100 °C are used to study the pulse width and temperature effects on fuel delivery at different stages.

The measured and predicted static flow rates and dynamic flows at different temperatures with a pulse of 2.5 ms are shown in Fig. 6. The predicted static flow rates in laminar flow and turbulent flow at 20 °C are 1.764 g/s and 1.778 g/s, respectively, which are very close to measured value of 1.74 g/s. In the temperature of 20 - 100 °C, the static flow rates predicted by laminar flow and turbulent flow show similar trend and are only different in less than 2%. This is because the maximum Reynolds number ranges from 2500 to 3700 at the injector exit which is not large enough to produce strong turbulence effects. Therefore, even the flow is not totally laminar, the flow rate predicted by the laminar flow is more accurate than that by turbulent flow. In view of this, the laminar flow approach is employed for the following numerical analysis. As the temperature moves higher, the predicted static flow rate in laminar flow increases to the maximum value of 1.775 g/s at 80 °C and drops to 1.754 g/s at 100 °C. This is the result of a decrease in both viscosity and density of the test fluid (M-10) at a high temperature. Lower viscosity tends to increase the flow velocity, whereas the lower density reduces the mass flow rate. The effects of the viscosity on mass flow rate are greater than that of the density before 80 °C. Though the predicted dynamic flow is 9 - 14.5% higher than the measured data, the overall trends of these two are very similar which increase with temperature before 80 °C and decrease after 80 °C. Compared to the test data, this computational analysis predicted reasonably good results.

The predicted mass flow rate during fuel injection at various pulse widths and a temperature of 20 °C is shown in Fig. 7. In the opening cycle, the flow rate goes up as the valve gradually opens. The flow rate at the fully open region increases slightly with the pulse width. At a long pulse width of 7.5 ms, the curve of transient flow rate at the fully open region finally merges to that of static flow rate. Note that the general shape of the mass flow rate curves is similar to that of needle lift which directly controls the fuel delivery. The dynamic flow

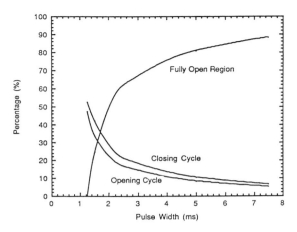

Fig. 9   Percentage of Dynamic Flow at Various Stages of Fuel Injection with Temperature of 20 °C

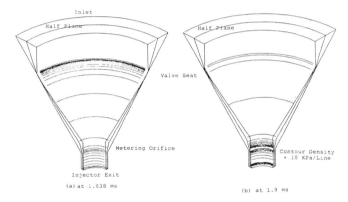

Fig. 10   Pressure Distribution at the Half Plane of Flow Passage with Temperature of 20 °C
(a) at 1.038 ms with Needle Lift of 0.015 mm
(b) at 1.9 ms with Needle Lift of 0.06 mm

Table 1   Effect of Temperature on Dynamic Flow at Different Stages with a Pulse Width of 2.5 ms

Temperature (C)	Dynamic Flow (mg/pulse)	Percentange (%) of Dynamic Flow at		
		Opening Cycle	Fully Open	Closing Cycle
20	3.467	16.82	62.0	21.18
40	3.503	17.09	61.77	21.14
60	3.568	17.23	61.57	21.20
80	3.585	17.28	61.50	21.22
100	3.549	17.48	61.27	21.25

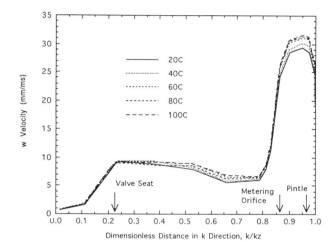

Fig. 11   Distribution of w-Velocity at 1.9 ms with Various Temperature and Pulse Width of 2.5 ms

because the smallest flow passage is now at the orifice and a large pressure head has been converted to the kinetic energy.

Table 1 shows the temperature effect on the flow delivered at different stages of fuel injection at a pulse width of 2.5 ms. A significant 40% of the fuel is delivered in the opening cycle (i.e., 16.82 - 17.48%) and closing cycle (i.e., 21.14 - 21.25%). An increase in temperature tends to increase the percentages of flow delivered at the opening and closing cycles. This is because at lower fluid viscosity the fuel is easier to move through narrow valve passages at the opening and closing cycles. As the temperature increases from 20 °C to 100 °C, the dynamic flow at the opening and closing cycles increase by 0.66% and 0.07%, respectively.

Figure 11 shows the temperature effect on spatial distribution of the w_k-velocity (velocity of the k component in the BFC) at the half plane of the flow passage at 1.9 ms. The injector inlet corresponds to the origin (i.e., 0) in the dimensionless unit of k/k_z and the injector exit to 1 in k/k_z. The w_k-velocity increases to the first peak at the valve seat, and then drops a little until in the vicinity of the metering orifice, and finally increases drastically to the maximum value of 29 - 32 m/s. Near the injector exit, the w_k-velocity drops a little because of the sudden expanded flow passage created by the pintle. The flow velocity is maximal at the temperature of 100 °C as a result of lowest fluid viscosity. When the needle is fully open at 1.9 ms, the flow velocities at the injector exit at temperature of 20, 40, 60, 80, 100 °C are 18.9, 19.3, 20.0, 20.5, 20.6 mm/ms which result in the maximum discharge coefficient (i.e., the ratio of the actual flow rate to theoretical flow rate) of 0.696, 0.707, 0.724, 0.735, and 0.732, respectively. These values are close to those predicted by the empirical equation for a plain orifice atomizer proposed by Nacayama (1961),

$$C_D = \frac{Re^{5/6}}{17.11\, l_o/d_o + 1.65\, Re^{0.8}} \quad (16)$$

where C_D is discharge coefficient at high Reynolds number, and l_o and d_o are the orifice length and diameter, respectively. From Eq.(16), one can calculate that for Re = 2500 - 3700 with $l_o/d_o = 4$,

at different pulse widths with a temperature of 20 °C is shown in Fig. 8. Linearity between the dynamic flow and pulse width can be observed at 1.5 - 7.5 ms with a slop of 0.433 mg/ms. The dynamic flow decreases quickly when the pulse width is smaller than 1.25 ms. This low-end nonlinearity is contributed to that a large amount of fuel is delivered at the opening and closing stages. Compared with the test results, the numerical data predict good general trend and values. From Fig. 9, one can see that at short pulse widths (< 2.5 ms), the percentage of dynamic flow delivered in the fully open region increases very quickly with the pulse width. About 50% of the fuel is injected at the fully open region when the pulse width is 2.0 ms at 20 °C, but about 89% of the fuel at a longer pulse of 7.5 ms. Also, even the valve opening time ($\Delta t = 0.55$ ms) and closing time ($\Delta t = 0.58$ ms) are almost the same, more fuel is delivered in the closing cycle. This is because at the beginning of the opening cycle, the pressure wave takes certain time to move to the injector exit in order to drive the flow, but at the closing cycle, the flowfield responded quickly to the needle movement.

Figure 10 (a) and (b) show pressure distributions at the half plane of the flow passage (i.e., $j = 0.5\,j_y$) at a temperature of 20 °C in an early stage (1.038 ms) and in the fully open stage (1.0 ms), respectively. At 1.0375 ms with the needle lift of 0.015 mm, most of the pressure gradient occurs at the valve seat because of the narrowest flow passage. As the time goes to 1.9 ms and the needle lift is 0.06 mm, the larger pressure gradient region moves to the metering orifice the discharge coefficient ranges from 0.729 to 0.754 which are slightly larger than predicted values. This is because the pintle injector has a divergent exit which results in a larger exit hydraulic diameter and a lower ratio of l_o/d_o. However, the basic functions of the metering orifice in pintle injector and the plain orifice in atomizer are very similar according to the close values in discharge coefficient.

CONCLUSIONS

Based on the analysis, the following conclusions can be drawn.

(1) At a temperature range of 20 - 100 °C, both static flow rate and dynamic flow are maximal at the temperature of 80 °C with a discharge coefficient of 0.735.

(2) At a pulse width smaller than 2.5 ms, the percentage of the fuel delivered at the fully open region increases rapidly with the pulse width.

(3) The temperature tends to increase the percentages of fuel delivered at the opening and closing cycles. More flow is delivered at the closing cycle than at the opening cycle.

(4) Pressure and w_k-velocity show small variation in the region between the valve seat and the metering orifice when the injector is fully open.

NOMENCLATURE

A_e	east surface area of a grid
C_D	discharge coefficient (i.e., ratio of actual flow rate to theoretical flow rate)
d_o	orifice diameter
F_u, G_u, H_u	local geometry factors in i, j, k directions
i, j, k	components in BFC (body fitted coordinates)
i_x	location of the symmetrical plane in BFC coordinates
j_y	location of the needle wall in BFC coordinates
k_v	location of the valve seat in BFC coordinates
k_z	location of the injector exit in BFC coordinates
l_o	orifice length
p	referenced pressure with respect to injector exit pressure
p_f	pressure at the fuel rail
r, θ, z	cylindrical coordinates
u, v, w	velocity components in Cartesian coordinates
u_i, v_j, w_k	velocity components in BFC coordinates
u_e, v_n, w_h	velocities at east, north, and high surfaces of a grid
V	vector form velocity
w_d	needle lift velocity in cylindrical coordinates
u_e^*	mass-flux velocity at east surface of a grid in BFC coordinates
α	inclination angle of needle surface in cylindrical coordinates
ρ_e	density at east surface of a grid
μ	dynamic viscosity

REFERENCES

Andrighetti, J. and D. Gallup, 1987, "Design-Development of The Lucas CAV Multipoint Gasoline Injector," SAE Paper 870127, presented at the International Congress and Exposition, Detroit, Michigan, February 23-27.

Greiner, M, P. Romann, and U. Steinbrenner, 1987, "BOSCH Fuel Injectors - New Developments," SAE Paper 870124, presented at the International Congress and Exposition, Detroit, Michigan, February 23-27.

Heywood, J.B., 1988, *Internal Combustion Fundamentals*, Chapter 7, McGraw-Hill, New York.

Hinze, J.O., 1975, *Turbulence*, McGraw-Hill, New York.

Launder, B.E. and D.B. Spalding, 1972, *Mathematical Models of Turbulence*, Academic Press, New York.

Nakayama, Y., 1961, "Action of the Fluid in the Air Micrometer: First Report, Characteristics of Small Diameter Nozzle and Orifice," *Bull. Japan Soc. Mech. Eng.*, vol. 4, pp. 516-524.

Okamoto, Y., N. Arai, K̄. Nakagawa, T. Kosuge, and T. Atago, 1992, "Atomization Characteristics of Two-Stream Injector for 4-Valve Engines," SAE Paper 920705, presented at the International Congress and Exposition, Detroit, Michigan, February 24-28.

Rosten, H. and D.B. Spalding, *The PHOENICS Equations*, CHAM TR/99, 1986, CHAM Limited, London, England, October.

SAE, 1992, "Gasoline Fuel Injector - SAE J1832 NOV89," SAE Handbook, Vol.3, pp. 24.246 - 24.262, Society of Automotive Engineers, Warrendale PA.

Spalding, D.B., 1989, *The PHOENICS Beginner's Guide*, CHAM TR/100, CHAM Limited, London, England, December.

Sugiura, S., T. Yamada, T. Inoue, K. Morinishi, and N. Satofuka, 1990, "Numerical Analysis of Flow in the Induction System of an Internal Combustion Engine - Multi-Dimensional Calculation Using a New Method of Lines," SAE Paper 900255, presented at the International Congress and Exposition, Detroit, Michigan, February 26-March 2.

ICE-Vol.18, New Developments in Off-Highway Engines
ASME 1992

PREDICTION TECHNIQUE FOR STRESS AND VIBRATION
OF NONLINEAR SUPPORTED, ROTATING CRANKSHAFTS

H.-H. Priebsch, J. Affenzeller, and S. Gran
AVL List GmbH
Graz, Austria

ABSTRACT

Design analyses of crankshafts including bearings, is necessary for both the layout of new engines and the modification of existing engines (increased power output, etc.). To improve the existing calculation systems for crankshafts and bearings, AVL has developed a new method. This method enables the coupled vibrations in the torsional, bending and axial directions, including gyroscopic effects, to be analysed.

For the simulation of multibearing effects the bearing models consider both the hydrodynamic oilfilm and the stiffness of the bearing structure. The calculation of forced vibrations is carried out using the gas and mass forces acting upon the rotating crankshaft.

Comparisons of calculated to measured results demonstrates the accuracy of this calculation technique. The method can be used for passenger car, truck and medium speed engines. In this paper examples of truck and passenger car engine applications confirm the additional possibilities for the estimation of crankshaft dynamics. Also the improvement of the results obtained from the new technique compared with those from classical calculation methods is described.

1 INTRODUCTION

The success of a new engine design is dependent on the durability, which is mainly influenced by the behaviour of the crankshaft and the bearings. Therefore great effort in simulation works is required for analyses of engine construction during the design stage. This is valid for both the layout of new engines and the improvement of existing engines. The high level of detailed refinement of engine design requires sophisticated tools for calculation and simulation systems. This is particulary true of a crankshaft and its bearings, for which a number of calculation methods exist. Nevertheless for modern engine design, optimisation problems occur due to some assumptions used in traditional calculation methods conflicting with current design demands.Without doubt improvement of engines using simulation and prediction techniques becomes more effective, the closer the calculation methods are to the real physical events.

For the analyses of bending stresses of crankshafts the common method assumes a statically determined supports [1]. Within this method gas and mass forces due to adjacent cranks only act on each bearing and crankshaft journal respectively. The torsinal vibrations are often calculated using a crankshaft model reduced to a plain shaft vibrating under moments due to gas and mass forces [2]. These methods include a decoupling of torsional and bending vibrations. The main disadvantage of this assumption is that a model created in this manner also neglects the coupling of bending and longitudinal vibrations. Most of these calculation methods use the influence of the engine structure known from experience, for the primary approximation.

Other common calculation methods for crankshafts use mean bearing stiffnesses [3,4,5]. This assumption is acceptable in the first approximation for torsional vibrations only, but not for bending vibrations. Also effects of the oilfilm nonlinearity leading to stiffness increases by a factor of 1000, are usually neglected for a multibearing crankshaft.

Bending modes and frequences may be greatly influenced by the stiffnesses of the bearing structure and the main bearing oil film. The frequencies of the bending modes may be changed by 30% or more when the actual stiffness of the bearings are considered.

For this reason AVL has developed the calculation system "KWDYN" to be able to predict crankshaft dynamics considering a multibearing rotating crankshaft. For these calculations a matrix method is used modelling the different parts of the crankshaft, the bearings, and the influence of the stiffness and damping of the engine structure. Matrix modeling methods include both, the global motion and deformation. This fact was found to be an advantage compared with Finite Element Methods (FEM) [6]. When using FEM for the calculation of forced vibrations the equations for the global motion of the model (rotation of the crankshaft) have to be build up for each element separately.

2 CALCULATION PROCEDURE

To calculate the coupled bending, torsional and longitudinal vibrations of crankshafts with regard to their non-linear bearings and rotation, an efficient procedure has been developed. This procedure allows to determine the motions of chainlike rotating structures build from massless beam and rigid single masses under non-linear bearing conditions.

To achieve this, the classical transfer matrix method [16] was modified. This matrix method offers the advantage to calculate in one run the small vibrational motion of the elastic structure as well as the big rigid-body motion. Furthermore, we assume that the chainlike structure may carry out large translations and rotations, its elastical distortion, however, will be small and can be described by means of the linear theory of elasticity.

The modified transfer matrix method will need only two different elements: a beam element and a mass element.

Nomenclature

$$\frac{d(\ )}{dt} \equiv (\dot{\ }) \quad ; \quad \frac{d^2(\ )}{dt^2} \equiv (\ddot{\ })$$

m	mass of a element
t	time
$\mathbf{e}_i$	(1×3) unit vectors of the inertial coordinate system
$\mathbf{e}'_i$	(1×3) unit vectors - definition of the structure's orientation
$\mathbf{x}$	(1×3) displacement vector (transversal)
$\mathbf{z}$	(1×13) state vector
$\mathbf{E}$	(6×6) or (3×3) unit matrix
$\mathbf{M}$	(1×3) vector of cutting moments
$\mathbf{M}_a$	(1×3) vector of excitation moments
$\mathbf{Q}$	(1×3) vector of cutting forces
$\mathbf{Q}_a$	(1×3) vector of excitation forces
$\mathbf{R}$	(3×3) transformation matrix
$\mathbf{U}$	(13×13) transfer matrix
$\boldsymbol{\varphi}$	(1×3) vector of angular displacement
$\boldsymbol{\omega}$	(1×3) vector of angular velocity of the mass element
$\boldsymbol{\Omega}$	(1×3) vector of reference angular velocity (structure)

2.1 Mass Element

In a general form, the equations of motion of a rigid body (element number i) freed from the chainlike structure will read as follows in the inertial system.

$$m^{(i)}\ddot{\mathbf{x}}^{(i)} = \mathbf{Q}^{(i)} - \mathbf{Q}^{(i-1)} + \mathbf{Q}_a^{(i)}$$
$$\mathbf{I}^{(i)}\cdot\dot{\boldsymbol{\omega}}^{(i)} + \dot{\mathbf{I}}^{(i)}\cdot\boldsymbol{\omega}^{(i)} = \mathbf{M}^{(i)} - \mathbf{M}^{(i-1)} + \mathbf{M}_a^{(i)} \tag{1}$$

We assumed that at the positive boundary of cutting (i) the cutting forces and moments will be written positively in the positive direction of the coordinates and at a negative boundary of cutting ($i-1$) pointing to the negative direction of coordinates. As boundary of cutting (i), the boundary between the elements with the numbers (i) and ($i-1$) is to be understood.

The entire elastic structure's orientation is determined by the three angles α, β and γ. The actual position of the freed rigid-body will

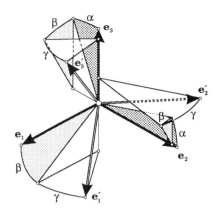

Fig. 1: Definition of the angles α, β and γ - Transformationsmatrix $\mathbf{R}$. - Definition of the structure's orientation.

then be determined by the small angles $\varphi_1^{(i)}$, $\varphi_2^{(i)}$ and $\varphi_3^{(i)}$, which will be treated as infinitesimal angles. The structure's orientation needs to be determined by approximation only. There is only one restriction, namely, that equation (2) always must be fulfilled for a structure consisting of n elements. Up to this limit the angels $\varphi_j^{(i)}$ can be considered as infinitesimal angels with sufficient accuracy.

$$\varphi_j^{(i)} \leq \varphi_{max} \approx 0,1 \stackrel{\wedge}{\approx} 6° \quad j = 1,2,3 \quad i = 1,2,...,n \tag{2}$$

Thus, the freed body's angular velocity $\boldsymbol{\omega}$ can be divided into the reference angular velocity $\boldsymbol{\Omega}$ and in a small share $\dot{\boldsymbol{\varphi}}$ which is generated by the distortion of this body with reference to the entire elastic structure's orientation defined by α, β and γ.

$$\boldsymbol{\omega} = \boldsymbol{\Omega} + \dot{\boldsymbol{\varphi}} \tag{3}$$

The reference angular velocity $\boldsymbol{\Omega}$ - which is valid for the total structure - needs not be determined by integration, but can be found by differentiation. The angles α, β and γ refer to the inertial system and define a transformation matrix $\mathbf{R}$, which transforms the coordinates of any tensors from the initial position into the actual orientation in the inertial system. The reference angular velocity $\boldsymbol{\Omega}$ may be given explicitly as the transformation matrix $\mathbf{R}$ always will be available in formula.

$$\boldsymbol{\Omega} = \dot{\mathbf{R}}\cdot\mathbf{R}^T \tag{4a}$$

By specially selecting α, β and γ (fig. 1) you may achieve that the simple relation (4b) will apply.

$$\boldsymbol{\Omega} = \begin{bmatrix} \dot{\alpha} & \dot{\beta} & \dot{\gamma} \end{bmatrix}^T \quad ; \quad \dot{\boldsymbol{\Omega}} = \begin{bmatrix} \ddot{\alpha} & \ddot{\beta} & \ddot{\gamma} \end{bmatrix}^T \tag{4b}$$

There we define that the system of coordinates ($\mathbf{e}'_1$, $\mathbf{e}'_2$, $\mathbf{e}'_3$) belonging to the reference orientation of the entire structure will be generated by subsequent rotations round the angles α, β und γ round the $\mathbf{e}_1$-, $\mathbf{e}_2$- and $\mathbf{e}_3$- axis of the inertial system. The rotations are shown in fig. 1

The special transformation matrix $\mathbf{R}$ belonging to it will read as follows:

$$\mathbf{R} = \begin{bmatrix} \cos\beta\cdot\cos\gamma & -\cos\alpha\cdot\sin\gamma + \sin\alpha\cdot\sin\beta\cdot\cos\gamma & \sin\alpha\cdot\sin\gamma + \cos\alpha\cdot\sin\beta\cdot\cos\gamma \\ \cos\beta\cdot\sin\gamma & \cos\alpha\cdot\cos\gamma + \sin\alpha\cdot\sin\beta\cdot\sin\gamma & -\sin\alpha\cdot\cos\gamma + \cos\alpha\cdot\sin\beta\cdot\sin\gamma \\ -\sin\beta & \sin\alpha\cdot\cos\beta & \cos\alpha\cdot\cos\beta \end{bmatrix} \quad (5)$$

The inertia tensor of the freed body in the entire elastic structure's orientation defined by $\mathbf{R}$ can be written as follows:

$$\underline{\mathbf{I}}^{(i)} = \mathbf{R}\cdot\mathbf{I}_0^{(i)}\cdot\mathbf{R}^{\mathrm{T}} . \quad (6)$$

$\mathbf{I}_0^{(i)}$ designates the body's inertia tensor belonging to $\alpha = \beta = \gamma = 0$. The actual inertia tensor measured in the inertial system can be determined with (7) in an extremely good, non-linear approximation.

$$\mathbf{I}^{(i)} = \underline{\mathbf{I}}^{(i)} + \mathbf{A}_\varphi^{(i)}\cdot\underline{\mathbf{I}}^{(i)} - \underline{\mathbf{I}}^{(i)}\cdot\mathbf{A}_\varphi^{(i)} ; \qquad \mathbf{A}_\varphi^{(i)} = \begin{bmatrix} 0 & -\varphi_3 & \varphi_2 \\ \varphi_3 & 0 & -\varphi_1 \\ -\varphi_2 & \varphi_1 & 0 \end{bmatrix}^{(i)} \quad (7)$$

With this, the first derivation in time of the inertia tensor appearing in the dynamical equations (1), can be laid down as follows:

$$\dot{\mathbf{I}}^{(i)} = \mathbf{A}_\omega^{(i)}\cdot\mathbf{I}^{(i)} - \mathbf{I}^{(i)}\cdot\mathbf{A}_\omega^{(i)} ; \qquad \mathbf{A}_\omega^{(i)} = \begin{bmatrix} 0 & \dot\gamma - \dot\varphi_3 & \dot\beta + \dot\varphi_2 \\ \dot\gamma + \dot\varphi_3 & 0 & \dot\alpha - \dot\varphi_1 \\ \dot\beta - \dot\varphi_2 & \dot\alpha + \dot\varphi_1 & 0 \end{bmatrix}^{(i)} \quad (8)$$

In a general form, an implicit direct integration statement will be brought in the dynamical equations (1) to partly eliminate the derivations in time. The basic idea of these integration methods is to express by approximation the derivations in time of a function $f(t)$ at a time t through the function to the time itself and through values of this function dated back in time $f(t-\Delta t)$, $f(t-2\Delta t)$, ... or through the derivations in time of this function at these times. All direct implicit methods of integration can be traced back to the following formal form:

$$\dot{f}(t) = \mu(\Delta t)\cdot f(t) - \tau_f ; \qquad \ddot{f}(t) = \xi(\Delta t)\cdot f(t) - \eta_f \quad (9)$$

The parameters μ and ξ only depend on the time step size Δt and the functions τ_f and η_f at the actual time t always are known. Appendix I gives the integration parameters μ and ξ as well as the functions τ_f and η_f for the most common methods.

If a vector, the so-called state vector $\mathbf{z}$, will be defined according to equation (19),

$$\mathbf{z} = [\mathbf{v} \quad \mathbf{p} \quad 1]^{\mathrm{T}} \quad ; \begin{cases} \mathbf{v} = [x_1 \; x_2 \; x_3 \; \varphi_1 \; \varphi_2 \; \varphi_3] \\ \mathbf{p} = [Q_1 \; Q_2 \; Q_3 \; M_1 \; M_2 \; M_3] \end{cases} \quad (10)$$

the equations (1) can be written in the form (11) by means of the integration statement with regard to $\mathbf{v}^{(i-1)} \equiv \mathbf{v}^{(i)}$ and the equations (3) to (8).

$$\mathbf{z}^{(i)} = \mathbf{U}^{(i)}\cdot\mathbf{z}^{(i-1)} \quad (11)$$

Here, $\mathbf{U}^{(i)}$ represents the modified transfer matrix of the general mass element in the form:

$$\mathbf{U}^{(i)} = \begin{bmatrix} \mathbf{E} & \mathbf{0} & \mathbf{0} \\ \mathbf{U}_{pv} & \mathbf{E} & \mathbf{u}_p \\ \mathbf{0} & \mathbf{0} & 1 \end{bmatrix}^{(i)} \quad (12a)$$

The partial matrix $\mathbf{U}_{pv}^{(i)}$ and the inhomogeneous part of the set of equations $\mathbf{u}_p^{(i)}$ will read as follows:

$$\mathbf{U}_{pv}^{(i)} = \begin{bmatrix} m\xi\mathbf{E} & \mathbf{0} \\ \mathbf{0} & \mu\dot{\mathbf{I}} + \xi\mathbf{I} \end{bmatrix}^{(i)} \quad (12b)$$

$$\mathbf{u}_p^{(i)} = \begin{bmatrix} \mathbf{Q}_a + m\eta_x \\ \mathbf{M}_a + \dot{\mathbf{I}}\cdot(\tau_\varphi - \mathbf{\Omega}) + \mathbf{I}\cdot(\eta_\varphi - \dot{\mathbf{\Omega}}) \end{bmatrix}^{(i)} \quad (12c)$$

In the transfer matrix, the magnitudes $\mathbf{I}$ and $\dot{\mathbf{I}}$ depending on $\boldsymbol{\varphi}$ and $\dot{\boldsymbol{\varphi}}$ will appear. However, this will create no problems, as the structure will be born non-linearly anyway. The non-linear hydrodynamical reaction forces of the bearing are replaced by external forces. They will be corrected iteratively till they have been predetermined accurately. Simultaneously, $\mathbf{I}$ and $\dot{\mathbf{I}}$ will be adapted. Due to the fact that $\mathbf{I}$ and $\dot{\mathbf{I}}$ appear as factors of $\boldsymbol{\varphi}$ in the transfer matrix, an implicit iteration algorithm is created which converges extremely safe. If the structure carries out large rotations only round one axis (e.g. crankshaft and connecting rod), the mass element transfer matrices can be brought in a form, where the elements of matrix $\mathbf{U}_{pv}^{(i)}$ are constant during the entire calculation and this iteration is no longer necessary. Appendix I gives an example of such a transfer matrix (flywheel element).

2.2 Beam element

By integration of the differential equation of the elastic line the transfer matrices of any beam elements arbitrarily orientated in space can be determined. If $\underline{\mathbf{U}}$ represents the beam transfer matrix in position $\alpha = \beta = \gamma = 0$, the required beam transfer matrix in the inertial system will read as follows (the transfer matrix for a straight beam is shown in appendix I):

$$\mathbf{U} = \mathbf{R}\cdot\underline{\mathbf{U}}\cdot\mathbf{R}^{\mathrm{T}} \quad (13)$$

2.3 Solution of the Equations of Motion

For a chainlike structure from n elements (beam and mass elements) all state vectors at the element's boundaries 1 to $n-1$ can be eliminated by simple matrix multiplications. First, all transfer matrices have to be generated according to the actually given orientation (α, β and γ) which is permissible according to equation (2). For this, the vectors $\boldsymbol{\eta}_x$, $\boldsymbol{\tau}_\varphi$ and $\boldsymbol{\eta}_\varphi$ appearing in the 13th column of the transfer matrices can be expressed by a corresponding integration statement (see appendix I) by means of already calculated displacements or velocities. At the start of a calculation, these vectors are defined by the initial conditions.

$$\mathbf{z}^{(n)} = \mathbf{U}^{(ges)} \cdot \mathbf{z}^{(0)}$$

$$\mathbf{U}^{(ges)} = \mathbf{U}^{(n)} \cdot \mathbf{U}^{(n-1)} \cdot \ldots \cdot \mathbf{U}^{(2)} \cdot \mathbf{U}^{(1)} = \prod_{i=n}^{1} \mathbf{U}^{(i)} \qquad (14)$$

The (13×13) total transfer matrix $\mathbf{U}^{(ges)}$ represents the linear connection between the two ends of the structure. From the 24 unknown, the unknown components of the state vectors $\mathbf{z}^{(0)}$ and $\mathbf{z}^{(n)}$, you have to eliminate 12 magnitudes to clearly solve the inhomogeneous (12×12) set of equations (14). At a free end of the structure $\mathbf{p} \equiv \mathbf{0}$ will apply for the cutting forces, at a fixed end $\mathbf{v} \equiv \mathbf{0}$ will apply. It is even possible to give a time curve of the boundary conditions. For further observation, we assume the two ends of the structure to be free which is the most frequently occuring case. With the partitionized total transfer matrix $\mathbf{U}^{(ges)}$

$$\mathbf{U}^{(ges)} = \begin{bmatrix} \mathbf{U}_{vv} & \mathbf{U}_{vp} & \mathbf{u}_v \\ \mathbf{U}_{pv} & \mathbf{U}_{pp} & \mathbf{u}_p \\ \mathbf{0} & \mathbf{0} & 1 \end{bmatrix}^{(ges)} \qquad (15)$$

the inhomogeneous (6×6) set of equations for the determination of the state vector's $\mathbf{z}^{(0)}$ unknown components will read

$$\mathbf{U}_{pv}^{(ges)} \cdot \mathbf{v}^{(0)} = -\mathbf{u}_p^{(ges)} \qquad (16)$$

Thus, the state vector is known and all displacements and cutting forces within the structure can be determined through continuous multiplication by the transfer matrices. With the distortions of all elements, condition (2) is to be controlled and the reference position of the structure to be corrected, if necessary. This can be effected by averaging these distortion over all elements.

$$\overline{\varphi}_j = \sum_{k=0}^{n} \varphi_j^{(k)} \qquad j = 1,2,3 \qquad (17)$$

The new orientation of the elastic structure then will be defined by the angles $\overline{\alpha}$, $\overline{\beta}$ and $\overline{\gamma}$ according to (18). In most cases, this adaptation will not be necessary within a time t as in practice the calculation is done with small computing time steps Δt and a sufficiently accurate orientation can be found for a new time $t + \Delta t$ by extrapolation.

$$\overline{\alpha} = \alpha + \overline{\varphi}_1, \quad \overline{\beta} = \beta + \overline{\varphi}_2, \quad \overline{\gamma} = \gamma + \overline{\varphi}_3 \qquad (18)$$

2.4 Macro Elements

When applying the method of transfer matrices, two measures are to be taken to eliminate the known numerical problems with structures containing a high number of elements. On the one hand, the element transfer matrices are made dimensionless with a mean element length, a mean moment of area of second order and the elasticity module of the structure and on the other hand so-called macro elements are introduced.

For this, the structure consisting of n elements will be divided up into q parts, the macro elements, with the permitted number of elements. From experience we know that 40 elements in dimensionless form can be processed without any numerical problems. The

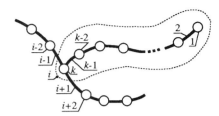

Fig. 2: Branched strukture

element boundaries of the macro elements have the numbers k_j, $j = 0, 1, \ldots, q$ with $k_0 = 0$ and $k_q = n$. The macro element transfer matrices $\mathbf{U}_M$ are determined by (19).

$$\mathbf{U}_M^{(j)} = \prod_{i=k_j}^{k_{j-1}+1} \mathbf{U}^{(i)} \qquad j = 1,2,\ldots,q \qquad (19)$$

These transfer matrices will be partitionized to

$$\mathbf{U}_M = \begin{bmatrix} \mathbf{U}_M & \tilde{\mathbf{u}}_M \\ \mathbf{0} & 1 \end{bmatrix}; \qquad \mathbf{z}_M = \begin{bmatrix} \tilde{\mathbf{z}}_M \\ 1 \end{bmatrix} \qquad (20)$$

With this, the following tapelike set of equations can be generated from which - after inclusion of the boundary conditions - the state vectors at the macro element boundaries can be determined with sufficient accuracy.

$$\begin{bmatrix} \tilde{\mathbf{U}}_M^{(1)} & -\mathbf{E} & & & \\ & \tilde{\mathbf{U}}_M^{(2)} & -\mathbf{E} & & \\ & & \cdot & \cdot & \\ & & & \cdot & \cdot \\ & & & & \tilde{\mathbf{U}}_M^{(q)} & -\mathbf{E} \end{bmatrix} \begin{bmatrix} \tilde{\mathbf{z}}_M^{(1)} \\ \tilde{\mathbf{z}}_M^{(2)} \\ \cdot \\ \cdot \\ \tilde{\mathbf{z}}_M^{(q)} \end{bmatrix} = \begin{bmatrix} \tilde{\mathbf{u}}_M^{(1)} \\ \tilde{\mathbf{u}}_M^{(2)} \\ \cdot \\ \cdot \\ \tilde{\mathbf{u}}_M^{(q)} \end{bmatrix} \qquad (21)$$

This set of equations can be solved most efficiently due to its bandlike form. Within the macro elements the state vectors are determined by continuous multiplications by means of the found state vectors of the macro element boundaries.

2.5 Branching

The so far assumed chainlike structure can easily be extended to chainlike structures with branchings (e.g. counter weight at a crankshaft). For this, the branched part will be reduced to an individual transfer matrix of the main structure. For the (corresponding to fig. 2) freed branched parts of the structure, the transfer matrix relation between the two ends of this structure will read as follows:

$$\begin{bmatrix} \overline{\mathbf{v}}^{(k)} \\ \overline{\mathbf{p}}^{(k)} \\ 1 \end{bmatrix} = \begin{bmatrix} \overline{\mathbf{U}}_{vv} & \overline{\mathbf{U}}_{vp} & \overline{\mathbf{u}}_v \\ \overline{\mathbf{U}}_{pv} & \overline{\mathbf{U}}_{pp} & \overline{\mathbf{u}}_p \\ \mathbf{0} & \mathbf{0} & 1 \end{bmatrix}^{(ges)} \begin{bmatrix} \overline{\mathbf{v}}^{(0)} \\ \mathbf{0} \\ 1 \end{bmatrix} \quad \text{mit} \quad \overline{\mathbf{U}}^{(ges)} = \prod_{j=k}^{1} \overline{\mathbf{U}}^{(j)} \quad (22)$$

From this equation changes of the cutting forces $\overline{\mathbf{p}}^{(k)}$ can be calculated due to the branched structural part (23).

$$\overline{\mathbf{p}}^{(k)} = \overline{\mathbf{U}}_{pv}^{(ges)} \cdot \overline{\mathbf{U}}_{vv}^{(ges)^{-1}} \cdot \overline{\mathbf{v}}^{(k)} + \overline{\mathbf{u}}_p^{(ges)} - \overline{\mathbf{U}}_{pv}^{(ges)} \cdot \overline{\mathbf{U}}_{vv}^{(ges)^{-1}} \cdot \overline{\mathbf{u}}_v^{(ges)} \quad (23)$$

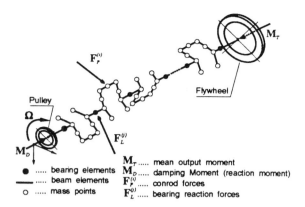

$$\mathbf{M}_T \dots \text{ mean output moment}$$

● ..... bearing elements $\mathbf{M}_T$ ..... mean output moment
— ..... beam elements $\mathbf{M}_D$ ..... damping Moment (reaction moment)
○ ..... mass points $\mathbf{F}_r^{(i)}$ ..... conrod forces
  $\mathbf{F}_L^{(j)}$ ..... bearing reaction forces

Fig. 3: Crankshaft model showing acting forces and moments for the transient multibearing calculation over 720 Deg. CA.

With this for the transfer matrix of the branched part reduced in the main structure with regard to $\mathbf{v}^{(i-1)} \equiv \mathbf{v}^{(i)}$ there will follow

$$\mathbf{U}^{(i)} = \begin{bmatrix} \mathbf{E} & \mathbf{0} & \mathbf{0} \\ \overline{\mathbf{U}}_{pv}^{(ges)} \cdot \overline{\mathbf{U}}_{vv}^{(ges)^{-1}} & \mathbf{E} & \overline{\mathbf{u}}_p^{(ges)} - \overline{\mathbf{U}}_{pv}^{(ges)} \cdot \overline{\mathbf{U}}_{vv}^{(ges)^{-1}} \cdot \overline{\mathbf{u}}_v^{(ges)} \\ \mathbf{0} & \mathbf{0} & 1 \end{bmatrix}^{(i)} \quad (24)$$

3 DESCRIPTION OF CALCULATION MODELS

3.1 Crankshaft

The matrices are built up starting with a 3D-Finite Element model of the shaft, cranks and webs (fig. 5). Using this model to calculate the deformations under unit loads and moments the elemental stiffness distribution along the crankshaft can be defined. Mass and inertia moments are reduced to the centres of gravity of selected nodal points of the crankshaft using CAD. A more detailed description of this step is given in [7,8]. After the definition of this centreline model of the crankshaft, mass, damping and stiffness values enable an elemental matrix model of the whole crankshaft to be generated. This model includes all dynamic properties of the crankshaft including gyroscopic effects, models for the flywheel and the damping moments at the pulley.

To check the accuracy of this calculation model, a comparison of calculated and measured free-free modes was carried out. The calculated modes considered interaction between bending, longitudinal, and torsional vibrations. These comparisons also helped to refine the model. Typical mean absolute differences of measured and calculated natural frequencies of up to 2000 Hz were less than 4 %.

Fig. 4 shows selected modes of a crankshaft of a 4 cyl. passenger car engine. The results are shown for first, second and third order bending modes of the crankshaft under free-free conditions. Differences in the natural frequencies are less than 4% (fig. 4).Differences of the mode shapes found from the comparison of measured and calculated results were negligible.

Modeshapes

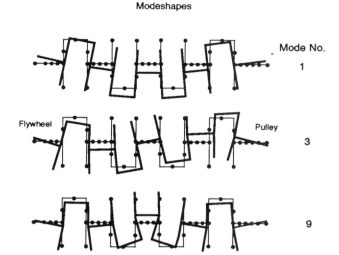

Comparison of frequencies - Hz

Mode No.	Measurement	Calculation
1	178	187
3	540	534
9	1631	1579

Fig. 4: Comparison of measured and calculated results of natural bending modes and frequencies for a 4 Cyl. passenger car engine

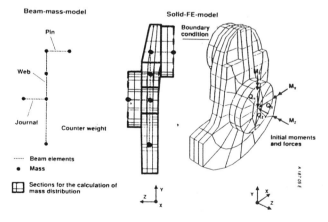

Fig. 5: Conversion of solid Finite Element model of one half crank throw to dynamic beam-mass model

To allow for calculation in the time domain with this crankshaft model, time integration terms for acceleration and velocity are automatically generated in the matrices. The time history calculation is necessary to consider nonlinear effects in the bearing oilfilms. For the forced vibration calculation, all external forces and moments acting at the cranks, and the crankshaft front and rear ends, are applied (fig.3). At the clutch side of the flywheel mean effective torque is considered.

3.2 Bearings

For the calculations described in this paper a hydrodynamic model was used for the main bearings. To check the accuracy of this bearing simulation technique a comparison of two principle models for oil film and bearings was performed. The first model enables the elastohydrodynamic effects in the bearing oil film, taking into account the interaction of pressure distribution and bore deformation of the bearing, to be considered. The second model uses a hydrodynamic calculation method for the oil film, and for stiffness and damping properties of the bearing structure. The principle differences of these two stiffness models can be seen from fig. 6.

In fig. 6 the hydrodynamic stiffness (C-HD), the stiffness of the bearing cap due to HD-pressure distribution (C-STR) and stiffness resulting from interaction of the elasticity of the structure and hydrodynamic oil film (C-EHD) are plotted. These calculations were carried out on the main bearing of a passenger car engine. The details of the calculation of C-EHD are described in [8].

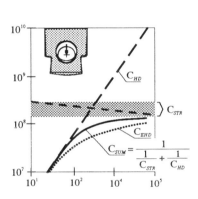

Fig. 6: Comparison of calculated stiffnesses of the main bearing structure and oil film, 2.3 l passenger car engine

STR structure due to HD press. distribution

HD hydrodynamic (nonelastic bore)

SUM elasto-hydrodynamic

$$C_{SUM} = \cfrac{1}{\cfrac{1}{C_{STR}} + \cfrac{1}{C_{HD}}}$$

Furthermore, fig. 6 shows that the resulting stiffness C_{SUM} of oil film and structure is closed to the oilfilm stiffness C_{HD} at low loads. At high loads the resulting stiffness approaches the stiffness of the bearing structure asymptotically.

At mean load forces and journal excentricities respectively the actual stiffness of oil film and bearing C_{EHD} is less than C_{SUM}, because of the different pressure distribution of the oil film. For the calculation of C_{EHD} the interaction between hydrodynamic the oil film pressure and the deformation of the bearing construction was taken into account.

Both bearing models, the hydrodynamic and the elastohydrodynamic simulate nonlinear properties of the oil film solving the Reynolds equation and the extended Reynolds equation respectively. For crankshaft stress and vibration analyses it could be found that hydrodynamic models are usually are sufficient accuracy. Therefore in the following examples of the application of KWDYN the hydrodynamic models are used together with structure stiffness of bearing construction.

Fig. 7 explains the bearing model. This model consists of hydrodynamic behaviour of the oil film due to journal motion and velocity

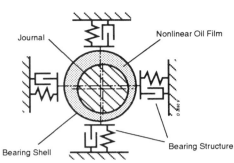

Fig. 7: Main bearing model for the calculation of transient multibearing crankshaft vibrations over 720 Deg. CA

in a rigid bore. Structure stiffness is applied using four pairs of springs and dampers. It is necessary to use different stiffnesses in different directions according to the structure behaviour. In this model a torsial spring is also included to simulate the interaction of crankshaft bending vibration and vibration of the bearing cap in the crankshaft axis direction.

In connecting these bearing models to the crankshaft model described above, a system of nonlinear equations is created, representing the dynamic behaviour of the vibration system and the interaction between rotating shaft and bearings (Basic analyses in [14]). This interaction between shaft vibration and oil film reaction force can only be solved iteratively. The vibration system has to be calculated in time domain using a step by step method of implicit-explicit type and the iteration has to be done in each time step. In this iterative process bearing reaction forces due to displacement and velocities of the journals in the bearings are calculated. These reaction forces together with initial forces and moments lead to a new set of journal velocities and crankshaft deformations. Convergency is controlled by the comparison of reaction forces calculated in two sequential iteration steps.

4 COMPARISON OF CALCULATION METHODS

A comparative analysis was carried out to find what influence different calculation methods had on the vibration and stress results. Models of crankshaft, oil film and bearings of a 6 cyl. truck diesel engine were used for these calculations as follows:

a.) Statically determined model with hydrodynamic oil film.

b.) Statically undetermined model with mean stiffness values for oilfilm and bearing structure.

c.) Dynamic multibearing (AVL's program KWDYN) including hydrodynamic oil film and structure stiffnesses.

Fig. 8 shows the calculation results for the bearing reaction forces of this engine's no. 7 main bearing under loads at 3000 rpm (overspeed). The statically determined method clearly leads to forces dominated by first order mass forces. The maximum amplitude of the bearing forces is about 65 % of those using dynamic multibearing method. The increase of maximum forces is dominated by dynamic effects, which are aused mainly by a natural mode vibration of the crankshaft in this example.

The statically undetermined method also leads to these dynamic effects, but with increased force amplitudes compared with the

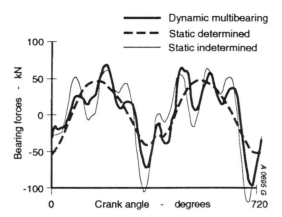

Legend:
——— Dynamic multibearing
– – – Static determined
——— Static indetermined

Fig. 8: Comparison of calculated main bearing forces using different calculation methods and models for a 6 cylinder truck diesel engine at 3000 rpm (overspeed)

results of KWDYN. Here one must recognise , that all calculation systems using the hydrodynamic models of the bearing oil film assume actual damping effects (version a. and c. in this comparison). Besides this effect a change of frequency can be seen for the statically undetermined system compared to results of KWDYN. This effect is caused by the different stiffness of the bearings, which is linear in the case of statically undetermined system and nonlinear in programm KWDYN. This fact also leads to different natural mode-shapes and frequencies in both calculation methods.

Fig. 9 shows calculation results for the journal orbital path in the bearing oil film obtained by different mathematical models. The comparative analysis was carried out for a 4 cylinder 4.5 l inline DI diesel engine at 4000 rpm (overspeed) for 4 and 8 counter weights. The hydrodynamic calculations were done for the bearing no. 3 with both a statically determined and a dynamic multibearing system. Having the same balance ratio for the whole crankshaft, the orbital path of the center bearing shows clearly effects of locally unbalanced rotating mass forces in this bearing due to the static determined model. The absolute motion of the journal is of course very different due to the elasticity of bearing structure.

5 MEASUREMENT AND CALCULATION RESULT AT THE FLYWHEEL

An important result of this calculation technique is the analysis of the flywheel motion, often discussed in literature [10,11,12,13,15].

To determine the accuracy of the calculation results in this view, comparison with measured results obtained on a fired engine were made. Fig. 10 and 11 show typical results of these comparisons for a 6 cyl. truck diesel engine at 1400 rpm and full load.

The measurement of the radial, axial and wobble vibrations of the flywheel were obtained from inductive displacement pick-ups. These pick-ups were mounted on the flywheel housing.

The comparisons show good agreement especially for the main vibration events at FTDC of cylinders no. 4,5 and 6 situated next to the flywheel. These vibrations are also influenced by the flywheel position, and the gear train width at the rear side of the engine respectively.

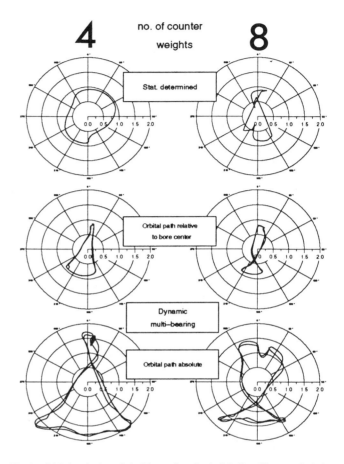

Fig. 9: Orbital path of crankshaft journal no. 3 calculated by different calculation methods for a 4 cylinder 4.5 l inline DI diesel engine at 4000 rpm, overspeed

Increased distances of the flywheel to the engine due to gear train width showed increased radial vibrations of the flywheel, particulary at overspeed. The wobbling motions were slightly reduced at the same engine condition.

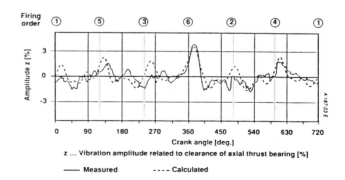

z ... Vibration amplitude related to clearance of axial thrust bearing [%]

——— Measured   - - - - Calculated

Fig. 10: Comparison of measured and calculated results of flywheel vibration in the crankshaft axis direction for a 6 cyl. truck diesel engine at 1300 rpm and full load

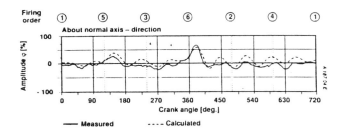

Fig. 11: Comparison of measured and calculated results of wobble motion of the flywheel for a 6 cyl. truck diesel engine at 1300 rpm and full load

Thus simulating different flywheel positions with this method, their influence on durability and acoustic4al problems can be analysed.

6 CRANKSHAFT VIBRATION

To illustrate the further advantages of this prediction technique. Fig. 12 shows calculated results of the crankshaft vibrations of a 2.3 l passenger car engine at 5000 rpm and full load. In this figure the deformed crankshaft centre line 10 Deg CA before, 5 Deg. CA and 20 Deg. CA after FTDC of each cylinder are ploted. Of course the ignition of a cylinder mainly causes deformation of the crankshaft close to that cylinder , but reactions due to vibrations of the crankshaft can be seen over the whole length.

Due to the elasticity of the main bearing structure journal vibrations may cause the original bearing clearence to be exceeded. It can be seen from fig. 12 that the maximum excentricity of journals in the bearings is about twice the original clearance of the bearings. Thus, this method can be used to analyse interaction effects of the stiffnesses of the crankshaft and engine structure.

The thrust bearing of an engine should be placed in respect to prevent from acoustical problems. Finding the optimal position for the axial thrust bearing may be of importance in the design stage of engines.

Bending Vibration in Liner Axial Direction over 720 Deg. CA

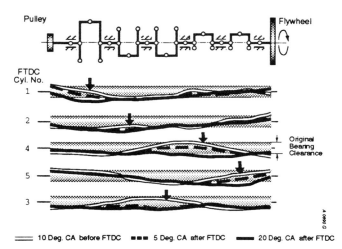

Fig. 12: Calculated bending vibration for the whole crankshaft over an engine cycle for 2.3 l passenger car engine at 5000 rpm and full load

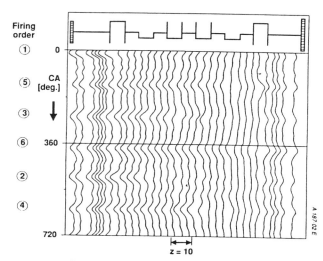

z ... Vibration amplitude related to clearance of axial thrust bearing

Fig. 13: Calculated axial vibration over crankshaft lenght for a 6 cyl. truck diesel engine at 1400 rpm and full load

Therefore postprocessing of KWDYN enables to plot axial vibrations over the whole crankshaft length. Fig. 13 shows these axial vibrations over the whole engine cycle for a 6 cyl. truck diesel engine at 1400 rpm and full load. Longitudinal vibration waves, caused by the ignition of each cylinder run between the crankshaft ends and result in maximum vibration amplitudes at the pulley. For this example the optimum position for the axial thrust bearing is at main bearing no. 6.

Finally moments and forces for the stress analysis can be ploted directly from the results of these calculation. The moments in torsional and bending directions as well as the shearing forces can be obtained for each beam of the crankshaft model.

Fig. 14 shows an example of analyses of principle design modifications of crankshaft and bearings of a 6 Cyl. truck diesel engine. In

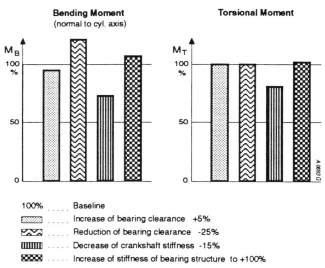

Fig. 14: Influence of principle design modifications on calculated stress moments of journal no. 7 for a 6 cyl. truck diesel engine at 3000 rpm (overspeed)

174

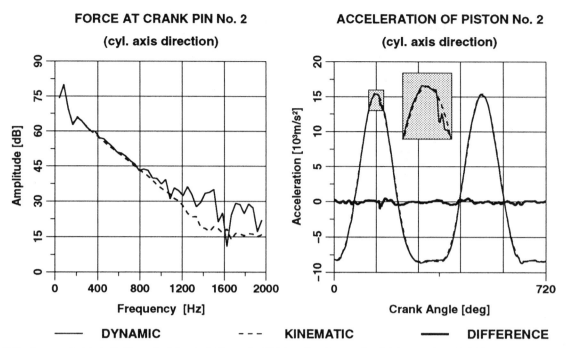

FORCE AT CRANK PIN No. 2

(cyl. axis direction)

ACCELERATION OF PISTON No. 2

(cyl. axis direction)

——— **DYNAMIC**   - - - **KINEMATIC**   ——— **DIFFERENCE**

Fig. 15: Excitation forces calculated with (dynamically) and w/o (kinematically) interactions to conrod and piston, crank no. 2, 2.4 l passenger car 5000 rpm, full load

this engine the maximum stresses were found at overspeed. The results confirm a dominating influence of the crankshaft stiffness. Bending moments of the crankshaft are also influenced by the change of bearing clearence, which of course means a change of bearing stiffness. A change of structure stiffness influences both the bending moments and the torsinal moments of the crankshaft.

7 VIBRATION OF CRANKTRAIN INCLUDING CONROD AND PISTON

In the running engine an interaction between the dynamics of piston and conrod on the one hand and the vibration of the rotating crankshaft on the other exists. The amount of the differences of the exitation forces on the crank between pure kinematic and dynamic crank loads is analysed for a 2.4 liter passenger car engine. For this analysis the models for the cranktrain and the bearings were extended by conrod and piston models. Each conrod is represented by two masses and a beam. The piston node is guided in cylinder axis direction at the cylinder center line. The coupling of the structure parts in the conrod bearings is performed by stiffness and damping forces. The forces (in the oilfilm) are nonlinear depending on relative movements and velocities. The nonlinear bearing characteristics are approximated by a precalculation of the hydrodynamic bearings.

The results of these calculations at 5000 rpm and full load are shown in fig. 15. The results indicate, that at frequencies lower than 800 Hz the excitation forces upon the crank no. 2 are negligibly affected by the dynamics of conrod and piston. At frequencies higher than 800 Hz the calculated dynamic forces increase compared to the kinematic ones. The differencies are 10 dB and more and thus showing considerable effects for noise excitation.

Also the resulting differences of kinematic and dynamic piston crankshaft, conrod and piston. Resulting differences between dynamic and kinematic piston motion are known from experiments and range up to 0.35 mm for this engine.

8 CONCLUSIONS

For the design analyses of the stress and vibration of crankshafts a newly developed calculation system is introduced in this paper. To demonstrate the advantage of this method for the simulation of the forced vibrations of crankshafts, results of applications for the analyses of passenger car and truck diesel engines were discussed. From these results and further investigations at AVL the following conclusions can be drawn:

- The described calculation system KWDYN enables the analysis of crankshaft vibrations in the view of:
  - bearing properties
  - flywheel position and gyroscopic effects
  - stiffness of engine construction
  - coupling of torsional, bending and axial vibration

- The results show influences of engine structure on bending and torsional vibrations and stress moments. These influences are depending on the ratio of both the crankshaft and the engine structure stiffnesses.

- Influences of a displaced flywheel position due to increased gear train width, increases the radial vibration mainly, but not its wobbling motions.

175

- The calculation program enables to consider also elements for conrods, pistons and their bearings.

9 ACKNOWLEDGEMENT

The authors wish to thank their colleagues at AVL for their contribution to the work described in this paper.

10 REFERENCES

[1] Cuppo E., Gaudio R.: Crankshaft bending stresses: Experimental Investigations and Calculation Methods, CIMAC Barcelona, 1975

[2] Gross W., Hussmann A.W.: Forces in the Main Bearings of Multicylinder Engines. SAE 660756 Chicago, 166

[3] Donath G., Seidemann H.: Auslegung von Dieselmotor- Kurbelwellen: Vergleich gemessener und gerechneter Spannungen - Teil 1 und 2 (Design of Diesel Engine Crankshafts: Comparison of Measured and Calculated Stresses - Part 1 and 2) MTZ 48 (1987) 6 and 11

[4] Parlevliet Th.: Modell zur Berechnung der erzwungenen Biege- und Torsionsschwingungen von Kurbelwellen unter Berücksichtigung der Ölverdrängungsdämpfung und Steifigkeit in den Grundlagern (Model for the Calculation of Forced Bending and Torsional Vibrations Considering Squeeze Damping and Stiffness in the Main Bearing Oilfilms), Theses, Mnchen, 1981

[5] Svoboda M., Ulrich H., Steinmller G.: Schwingungsberechnung bei der Entwicklung von Pkw-Motoren (Vibration Analysis Developing Passenger Car Engines) VDI-Berichte Nr. 113 (1986)

[6] Gran S.: Berechnung der rotierender Kurbelwelle bei mehrfacher, nichtlinearer Gleitlagerung (Calculation of Rotating Cankshaft Considering Nonlinearity of their Slider Bearings), Dipl. Arbeit, TU Graz, 1989

[7] Priebsch H.H., Affenzeller J., Kuipers G.: Prediction Techniques of Vibration Transfer in Engines, C420-023, IMechE London, 1990

[8] Priebsch H.H., Affenzeller J., Kuipers G.: Structure Borne Noise Prediction Techniques, SAE 900019, Detroit 1990

[9] Schweiger W., Vollan A., Dirschmid W.: Zur Strukturdynamik von Kurbelwellen (Structure Dynamics of Crankshafts). VDI-Berichte Nr. 444 (1982)

[10] Yamashita T., Takahara M, Tsujimura A.: A Convenient Calculation Method on Bending Stress of Crankshaft. VDI-Berichte Nr. 370 (1980)

[11] Okamura H., Sogabe K.: Dynamic Stiffness Matrix Method for the three - Dimensional Analysis of Crankshaft Vibrations. C23/88 IMechE (1988)

[12] Yoshikawa K.: Vibration of Crankshafts at High Engine Speeds. FISITA, 1986

[13] Hodgetts D.: The Whirl Modes of Vibration of a Crankshaft. IMechE C216/76, 1976

[14] Priebsch, H.H.: Rotordynamik bei großer Unwucht mit Berücksichtigung nichtlinearer Feder- und Dämpfungseigenschaften der Lagerung (Rotordynamics with High Excenticity Ratios Considering Nonlinear Spring and Damping Charecteristics in the Bearings) Theses, Graz, 1980

[15] Heath A.R., McNamara P.M.: Crankshaft Stress Analysis - combination of Finite Element and Classical Analysis Techniques, Trans. ASME, J. of Eng. f. Gas Turbines and Power Plan A,vol. 112 July 1990

[16] Pestel E. C., Leckie F. A.: Matrix Methods in Elastomechanics; McGraw - Hill Book Company, 1963

[17] Bathe K.-J. Finite Elemente Methode. Springer Verlag, 1990

APPENDIX I

3-point differences method

$$\mu = {}^3/_{(2\Delta t)} \qquad \xi = {}^1/_{(\Delta t^2)}$$
$$\tau_f = -{}^{1}\!/_3 \cdot [-4f(t-\Delta t) + f(t-2\Delta t)] \tag{25a}$$
$$\eta_f = -\xi \cdot [-2f(t-\Delta t) + f(t-2\Delta t)]$$

HOUBOLD's method

$$\mu = {}^{11}\!/_{6\Delta t} \qquad \xi = {}^2\!/_{\Delta t^2}$$
$$\tau_f = -{}^{1}\!/_{11}[-18f(t-\Delta t) + 9f(t-2\Delta t) - 2f(t-3\Delta t)] \tag{25b}$$
$$\eta_f = -{}^{1}\!/_2[-5f(t-\Delta t) + 4f(t-2\Delta t) - f(t-3\Delta t)]$$

NEWMARK's method [17] ($\delta \geq {}^1\!/_2$, $\alpha \geq {}^1\!/_2({}^1\!/_4+\delta)^2$)

$$\mu \equiv a_1 = {}^\delta\!/_{\alpha\Delta t} \qquad \xi \equiv a_0 = {}^1\!/_{\alpha\Delta t^2}$$

$$\tau_f = a_1 \cdot f(t-\Delta t) + a_4 \cdot \dot{f}(t-\Delta t) + a_5 \cdot \ddot{f}(t-\Delta t)$$
$$\eta_f = a_0 \cdot f(t-\Delta t) + a_2 \cdot \dot{f}(t-\Delta t) + a_3 \cdot \ddot{f}(t-\Delta t) \tag{25c}$$

$$a_0 = {}^1\!/_{(\alpha\Delta t^2)} \qquad a_2 = {}^1\!/_{(\alpha\Delta t)} \qquad a_4 = {}^\delta\!/_\alpha - 1$$
$$a_1 = {}^\delta\!/_{(\alpha\Delta t)} \qquad a_3 = {}^1\!/_{(2\alpha)} - 1 \qquad a_5 = {}^{\Delta t}\!/_2 \, ({}^\delta\!/_\alpha - 2)$$

Example of a masselement transfer matrix - flywheel

Fig. 16 schows a flywheel mass element.

$$\mathbf{U} = \begin{bmatrix} \mathbf{E} & \mathbf{0} & \mathbf{0} \\ \mathbf{U}_{pv} & \mathbf{E} & \mathbf{u}_p \\ \mathbf{0} & \mathbf{0} & 1 \end{bmatrix} \tag{26}$$

$$\mathbf{U}_{pv} = \begin{bmatrix} m\xi & 0 & 0 & 0 & 0 & 0 \\ 0 & m\xi & 0 & 0 & 0 & 0 \\ 0 & 0 & m\xi & 0 & 0 & 0 \\ 0 & 0 & 0 & I_I\xi & 0 & 0 \\ 0 & 0 & 0 & 0 & I_{II}\xi & \Delta I\tau \\ 0 & 0 & 0 & 0 & -\Delta I\tau & I_{II}\xi \end{bmatrix} \tag{27a}$$

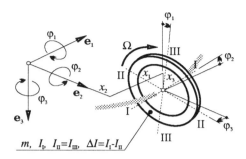

Fig. 16: mass element - flywheel

$$
\mathbf{u}_p = \begin{bmatrix} m\eta_{x_1} + Q_{1,a} \\ m\eta_{x_2} + Q_{2,a} \\ m\eta_{x_3} + Q_{3,a} \\ I_I\eta_{\varphi_1} + M_{1,a} \\ I_{II}\eta_{\varphi_2} + \Delta I\tau_{\varphi_3} + M_{2,a} \\ I_{II}\eta_{\varphi_3} - \Delta I\tau_{\varphi_2} + M_{3,a} \end{bmatrix}
\qquad (27b)
$$

Example of a beam transfer matrix

The transfer matrix for a straight uniform beam (fig. 17) with principal axis parallel orientated to the vektors of the $\mathbf{e}_i$- axis is given by (28).

$$
\underline{\mathbf{U}} = \begin{bmatrix} \mathbf{E} & \mathbf{B}_I & \mathbf{B}_{II} & -\mathbf{B}_{III}^T & \mathbf{b} \\ \mathbf{0} & \mathbf{E} & \mathbf{B}_{III} & \mathbf{B}_{IV} & \mathbf{0} \\ \mathbf{0} & \mathbf{0} & \mathbf{E} & \mathbf{0} & \mathbf{0} \\ \mathbf{0} & \mathbf{0} & \mathbf{B}_I & \mathbf{E} & \mathbf{0} \\ \mathbf{0} & \mathbf{0} & \mathbf{0} & \mathbf{0} & 1 \end{bmatrix}
\qquad (28)
$$

The associated partial (3×3) matrices are:

$$
\mathbf{B}_I = \begin{bmatrix} 0 & 0 & 0 \\ 0 & 0 & L \\ 0 & -L & 0 \end{bmatrix}
\qquad (29a)
$$

$$
\mathbf{B}_{II} = \begin{bmatrix} L/(EA) & 0 & 0 \\ 0 & -L^3/(6EI_3) + \kappa_2 L/(GA) & 0 \\ 0 & -L & -L^3/(6EI_2) + \kappa_3 L/(GA) \end{bmatrix}
\qquad (29b)
$$

$$
\mathbf{B}_{III} = \begin{bmatrix} 0 & 0 & 0 \\ 0 & 0 & L^2/(2EI_3) \\ 0 & -L^2/(2EI_2) & 0 \end{bmatrix}
\qquad (29c)
$$

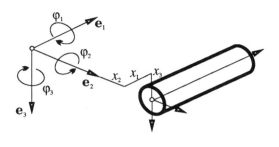

Fig. 17: Straight uniform beam element

$$
\mathbf{B}_{IV} = \begin{bmatrix} L/(GI_t) & 0 & 0 \\ 0 & L/(EI_2) & 0 \\ 0 & 0 & L/(EI_3) \end{bmatrix}
\qquad (29d)
$$

$$
\mathbf{b} = \begin{bmatrix} L \\ 0 \\ 0 \end{bmatrix}
\qquad (29e)
$$

E Modulus of elasticity of the beam material
G Thrust modulus of the beam material
L Lenght of the beam element
A Sectional area of the beam [m2]
I_2, I_3 2nd deg. moment of area round the $\mathbf{e}_2$, $\mathbf{e}_3$-axis
I_t Torsional moment of area of 2nd deg.

κ_2, κ_3 Thrust adjustment figure in $\mathbf{e}_2$, $\mathbf{e}_3$-direction

MANAGING ISO 9000 IMPLEMENTATION

William J. Lupear
Terrecorp, Incorporated
Clinton, Indiana

ABSTRACT

Many companies world wide are implementing and applying for registration for the ISO 9000 series of quality standards. This implementation has the same effect on an organization as any other process that is changing the methods, procedures, or the basic culture of the company. The management style necessary to make these changes successful is the same as that which is necessary for any other transformational process. If the correct management style is present, ISO 9000 implementation, and any other change being attempted, will be successful. Change can be accomplished without transformational management but the overall effect and the cost benefit will not be maximized.

This transformational management is the most important element. The technical aspects (the ISO 9000 hows and whys) are just a matter of someone within the organization learning them and developing a plan and system for management. The correct management style to achieve change is the most difficult. Strategies need to be clearly defined and methods developed in order to control the projects. Roadblocks need to be clearly identified and action planned to compensate for any shortcomings.

INTRODUCTION

In 1979 the ISO Technical Committee 176 was charted with the task to coordinate the international activities of quality systems. The objective was to insure good control practices and procedures for the mutual benefit of the supplier and customer. The result was the publishing of the ISO 9000 series of quality standards in 1987. These quality standards were first adopted by the European Community but have gained widespread acceptance world wide (51 countries). As these standards gain acceptability, more and more US companies are finding themselves in the position of changing and/or improving their quality system to comply with the ISO standards, especially if they deal in world markets. More customers are requiring that their suppliers have ISO 9000 series certification.

The actual activities associated with developing and implementing an ISO 9000 system have been well documented in numerous books and magazines. The scope of this paper is aimed at the management methods, strategies, and implementation procedures that I've experienced in the ISO chase.

MANAGEMENT METHOD

Management must use a method or style to insure change is possible. The type of transformational management proposed by Kim Cameron in his book Developing Management Skills(1991)[1] or by J. Juran in his book Managerial Breakthrough (1964)[2] is the

---

1 Cameron, Kim, Developing Management Skills, (New York: Harper-Collins, 1991).

2 Juran, J.M., Managerial Breakthrough, (New York: McGraw-Hill, 1964)

type of management style to which I am referring. This change process must be planned, developed, and then implemented. Most important of all, this change must have the commitment of all involved, including top management. No matter how skilled and knowledgeable an organization is on the subject of the ISO 9000 series of standards, it won't accept and achieve the results expected unless the proper management style and top management support exists.

Three criteria must exist before any implementation is started; top management must be committed and be involved in all phases of the process, a system to allow for change must exist within the organization, and there must be a total involvement of all the people in the process of change and their needs addressed.

Management commitment will show all that the change is important, necessary for the success of the company, and warrants the resources applied to the change process. Management must not only charter the ISO implementation, but must also be involved in the day to day activities.

The system that will allow change needs to be tailored to each organizational culture. There are general guidelines for this change process, however, the specific recipe will have to be developed by someone with the expertise to clearly identify the culture and what methods are needed. Many times this individual will be someone from outside the company.

The employees must be a part of this process and need to be involved in all aspects of this change process. The people must be treated as associates, not just employees. Their input should be sought when developing the planning and linkage to every day activities. They must have a thorough understanding of what the requirements of the ISO 9000 standards are, and how their specific responsibilities will be affected. It is managements responsibility to supply whatever is necessary to achieve this understanding is supplied. In addition to normal supervision, this could take the form of employee participation groups, formal or informal training, published procedures, published policies, corporate mission statements, and operating principles. It is establishing the "state of self control" for all employees. This concept explained in Managerial Breakthrough (1964)[3] and in the Quality Control Handbook (1974)[4] by J. M. Juran is; know what the objectives are, know whether they are being met, and the ability and authority to regulate performance in order to achieve the

objectives. The ability and authority to regulate is a very important part of this concept. Many times the authority is not truly given by management. This concept is basic but it applies to ISO implementation, establishing good working practices, Statistic Process Control, or any change or new addition to an organization.

STRATEGIES

A company involved in implementing a change such as ISO 9000 must develop a strategy for structuring this change, and monitoring and controlling the activities. The seven step process for formulating fundamental transformation to an ISO 9000 system should be followed. These steps are; recognize the need for change, initiating an internal study to assess organizational needs, decide to pursue the change, formulate a holistic quality policy adapting the ISO requirements, commit to support the policy by all personnel, plan and schedule how to implement the system, and initiate implemetation. How these individual steps are accomplished must be determined by each organization, however, I would like to suggest that existing project planning techniques can be used to accomplish these tasks.

Task Gantt, Resource Gantt, and PERT charts are just a few of the possible planning and control devices that can be used. A color coded project status board might be considered for an organization that require constant monitoring and updating. If this status board is prominately displayed, it has the added advantage of letting top management see if all of the tasks are on schedule. Special reporting will not be necessary if top management intervention is necessary.

As complex and mutifaceted as ISO 9000 implementation is, some type of monitoring system must be used to insure all aspects of the standards have been addressed. Many companies have based their supplier audits on the ISO standards. Theses audit forms are ideal to use as a guide in doing a self-audit in preparation for registration or a customer audit.

If an in-house expert on ISO 9000 does not exist, the

_____

3 Juran, p. 189.

4 Juran, J. M., Quality Control Handbook, (New York: McGraw-Hill, 1974), p. 2-13.

company can hire a consultant to help guide and steer the company. This is especially important thru the planning phase. In addition, the consultant can bear the brunt of any adverse reaction and take the heat if the implementation is not sucessful.

Another alternative is to develop an internal expert through seminars and ISO specific training. This is time consuming and will take a project manager type integrator to be successful. If this path is chosen, several very hard questions need to be answered, as the candidates for this type of responsibility are hard to find. If the company is progressive and is involved with many R&D projects, this individual can be very valuable. In the present day competitive markets the need for highly specialized functions and tight intergration between disciplines is critical. This integrator can fill that need. This person would also be ideal in the facilitation of employee involvement groups that would be used from the planning stages through implementation of the ISO 9000 standards.

ROADBLOCKS

After developing the proper controls and systems, the possible roadblocks to success should be addressed. There are many tools that can be used to help identify problems and plan countermeasures. I use a method similar to Failure Mode, Effects, and Criticality Analysis (FMECA) as outlined in MIL-STD-1629A[5] . This method is primarily for product design from concept through development but with modification it can be used for failures within a process/system implementation. The objective of my analysis is to identify possible failures with the implementation plan, identification of these failures as they occur, and countermeasures to eliminate or minimize them. This analysis should be performed throughout the planning and preliminary implementation. I have found that this is an ideal tool to use when involving the personnel affected by the ISO implementation. Their input is sought during the planning stage and brainstorming sessions can be used to suggest possible failure items on the FMECA list. This also has the added benefit of the personnel "buying-in" to the change.

Another tool to use is the fishbone diagram, also called cause and effect or Ishikawa diagrams. This is a very common tool used by Quality Control personnel but it can also be used to help dissect a problem and find the true cause of the problem. It is a very useful tool to be used by groups assigned the task of identifying and planning for corrective action to problems.

Choice of which methods used to accomplish the task of controlling problems encountered in the ISO 9000 implementation process is a matter of personal preference and which methods best match the company culture. It's very important, however, that this task be performed and not taken for granted. Problems will occur and they can have a negative impact on the sucess of the project.

IMPLEMENTATION

After the proper planning has been accomplished, it is a matter of following the task list with mid-course corrections. The task list should have all the tasks identified, prioritized, and a time element established for each task. Predecessor and successor tasks should be clearly identified and the sequence understood.

One of the first tasks should be the formulation of the Quality Manual. This manual must fulfill the requirements of the ISO standard, however, it should truly reflect the philosophy and policies of the company. All personnel, especially top management, should have input into the development of this manual. After all, everyone in the organization will have to live with this document. To many times someone will write a manual that will say all the right things, but it will only be lip service if everyone hasn't bought-in to the new system. At this point in the implementation process, the company should search it's corporate soul to identify what it wants to accomplish and does it truly want and need the ISO certificatiion and the system that goes with it.

The second task will be to document the operating procedures. These procedures will cover the who, what, when and where of the daily operations through-out the organization. At this point I would suggest that the individual departments put in writing, if it does not exist, the procedures that are presently being followed. These procedures can then be compared to the ISO standards to see if any modifications are needed. These modifications can then be documented and implemented into the day to day activities. If everyone was involved in the early phases of this process, they will fully understand why these changes are necessary. There will be very little resistance to the change.

_____

5 Department of Defense, <u>Procedure for Performing a Failure mode, Effects and Criticality Analysis</u>, <u>MIL-STD-1629A</u>, (Washington, DC: US Government Priniting Office, 1980).

The third task is to write the specific work instructions. This would include specific instructions for machines, gages, instruments, audit and inspection, etc.. If a company was to write these instructions for every work task, it could be a never ending process. Only those instructions directly related to the ISO standard need to be documented.

SUMMARY

If the right steps are taken early in the planning and developing phases, the actual implementation and certification will be comparatively smooth. Management needs to be aware and act on the necessary activities that will insure a successful implementation. If the company is rigid and reluctant to change, the ISO implementation will be a fight and the results will be wanting. The company has to not only know what the destination is, but how they are going to get there.

The ISO 9000 Standards have their strengths and weaknesses. The standard is to bring companies together under one umbrella, avoiding confusion, and building one common communication tool. Registration is becoming a necessity in some markets, and will surely strengthen a company's position in any market. As a criticism, I do not feel that the ISO 9000 Standards are enough of a quality effort for a company dealing in world markets and pursuing constant improvement. A company needs to stay competitive and constantly improve their systems, including quality. ISO is just a start towards Total Quality Management and Improvement. Many companies have excellent quality systems already in place and getting ISO certified is a fairly simple process for them. Others have weak or non-existent quality systems and it's a major struggle for them. Unfortunately, the companies with the poor quality systems are usually the ones with the cultures that have difficulty changing. If they understood the need for good quality systems, they would understand the purpose of the ISO 9000 Standards.

Understanding the concept of the ISO 9000 Standards and quality systems cannot be overemphasized. A company and all it's people must understand the necessity of world class quality in order to survive in today's markets. If they do not understand and implement, they will have a difficult time surviving.

The question of cost is always a consideration in any project or proposal. The cost of installing an ISO 9000 system will vary depending on what level of quality a company has achieved before ISO. For some companies it will be a major expenditure, for others, only the cost of the activities of the assessor performing the ISO accreditation. My question is, how can any company afford not to pursue a better quality system? ISO 9000 is a good start towards a world class system. A company can save the cost of the certification process if it's not necessary for their market, however, every company should be striving towards improvement if they want to stay competitive.

A company's quality objectives should be more ambitious than the ISO standards. The Malcolm Baldrige National Quality Award is a standard of business excellence in the United States. The evaluation criteria for this award is an excellent guide for business to follow in development of their quality systems. It addresses such issues as strategic quality planning, human resource development and management, information and analysis, and the very important issues of customer focus and satistaction. I would highly recommend that anyone interested in improving their system use a copy of the Baldrige Award Criteria to plan and steer their activities.

REFERENCES:

Cameron, K., 1991, "Developing Management Skills," Harper-Collins, New York, N.Y.

Department of Commerce, NIST, 1992, "Malcolm Baldrige National Quality Award Criteria," American Society for Quality Control, Milwaukee, WI.

Department of Defense, 1980, "Procedures for Performing a Failure Mode, Effects and Criticality Analysis, MIL-STD- 1629A," Navy Publishing and Printing Service, Philadelphia, PA.

Juran, J. M., 1964, "Managerial Breakthrough," McGraw-Hill, New York, N.Y.

Juran, J. M., 1974, "Quality Control Handbook," McGraw-Hill, New York, N.Y.

Lamprecht, J. L., 1991, "ISO 9000 Implementation Strategies," Quality, Vol. 30, No. 11, Hitchcock Publishing Co., Carol Stream, IL.

Meckstroth, D. J., 1992, "The European Community's New Approach to Regulation of Product Standards and Quality Assurance (ISO 9000): What it Means for U.S. Manufacturers," MAPI Economic Report ER-218, Manufacturers' Alliance for Productivity and Innovation, Washington, D.C.

Warchol, M. H. 1992, "Understanding ISO 9000: Part 1," Modern Casting, Vol. 82, No. 5, American Foundrymen's Society, Des Plaines, IL.

Wheeler, S., 1992, "Guidelines for Managing Organizational Change," Manufacturing Systems, Vol. 10, No. 6, Hitchcock Publishing Co., Carol Stream, IL.

ICE-Vol.18, New Developments in Off-Highway Engines
ASME 1992

FLEET-SITE MEASUREMENTS OF EXHAUST GAS EMISSIONS FROM URBAN BUSES

R. Bata, W. Wang, M. Gautam, Donald W. Lyons, N. Clark, M. Palmer, D. Ferguson, and S. Katragadda
Department of Mechanical and Aerospace Engineering
West Virginia University
Morgantown, West Virginia

ABSTRACT

In 1990, the Clean Air Act was signed into law giving four years for urban buses and eight years for urban trucks to comply with stringent regulations to control emissions from IC engines. In response, the Department of Energy (DOE) contracted West Virginia University to develop a Transportable Heavy Duty Vehicle Emissions Testing Laboratory to measure the levels of production of the regulated pollutants. The lab performs chassis dynamometer tests since this method produces more realistic emission values than engine dynamometer tests. In Houston, Texas, four urban transportation buses were tested over the Central Business District Cycle (CBD) under ambient conditions. Two of the buses were equipped with dual-fueled engines and were powered with No.2 diesel (D2)-liquified natural gas (LNG), clean diesel (CD)-LNG, D2/compressed natural gas (CNG), and CD-CNG. The other two had single-fueled engines and were powered with D2 and CD. The heavy duty test procedures of the Federal Register (code 40) were followed in all tests. Clean diesel produced less carbon monoxide (CO), nitrogen oxides (NO_x), and particulate matter (PM), but higher total hydrocarbons (HC) than D2. Dual-fuels containing CNG or LNG produced slightly less PM than D2 but higher HC and CO than both CD and D2. Among the dual-fuels D2-CNG produced the lowest levels of CO, HC, and PM.

INTRODUCTION

The transportation sector has been known to be a significant contributor to the urban air pollution problem. Many governments have considered the problem serious enough to implement strict emission regulation standards. These standards are being changed in an increasingly stringent manner since extensive research has been done on both the characterization of exhaust emissions and on the evaluation of the contribution of individual components to air pollution. To meet these standards, engine manufacturers responded by improving fuel metering devices, implementing closed-loop control systems, improving in-cylinder combustion and applying post combustion control technologies. As a result, a 92% reduction in pollutant emissions from vehicles has been achieved over the past twenty years (1). However, air quality has continued to decline in many urban areas despite the efforts to control pollutant emissions through implementing strict regulations. Another solution is being sought through using alternative fuels as well as reformulated conventional fuels.

Exhaust emissions from heavy duty engines (trucks and buses) are considered a significant source of ambient air pollution. Facilities with the necessary equipment for testing and measuring exhaust emissions are stationary and trucks and buses have to move to these facilities to be tested. To minimize the vehicles' time out of service and to encourage fleet owners to test their trucks and buses, the DOE contracted West Virginia University (WVU), in 1990, to build the first Transportable Heavy Duty Vehicle Emissions Testing Laboratory (2,3) to perform these tests at fleet sites.

The Laboratory measures mass concentrations of HC, CO, NO_x, carbon dioxide (CO_2) and PM. The present testing method is different from the emissions certification test since the latter utilizes an engine dynamometer, and emission rates are stated as grams per brake horsepower hour (g/Bhp-hr) while ours utilizes a chassis dynamometer test over a prescribed driving cycle, and the emission rates are determined in terms of grams per mile.

Findings show that engine certification tests could under estimate emissions contribution from these engines when they are installed in different vehicles with different drive train systems, (4,5). In order to create a database of emissions from heavy duty vehicles running on alternate and/or fossil fuels, based on the more realistic chassis dynamometer testing, the DOE and the Federal Transit Administration (FTA) have set up a 2-year testing plan for the lab. In the first year, eighty urban buses and trucks will be tested at different sites in the United States. Houston, Phoenix, Pittsburgh, New York City, San Francisco, Peoria, Tacoma, Parkersburg, WV, and many other locations are recommended as testing sites.

This paper presents the results of testing four buses in Houston, TX. The buses were equipped with single- and dual-fuel engines operating on D2, CD, LNG and CNG.

LABORATORY DESCRIPTION AND TEST PREPARATION

The laboratory, Figure 1, consists of three major mobile integrated systems: the chassis dynamometer system (power absorber units, flywheel units, and drive train units), the exhaust gas analysis system (bench analyzer, dilution tunnels, critical flow ventures, and blower), and a control and data acquisition system (2, 3). The emissions from the vehicle enter a constant volume sampler (CVS) full size stainless steel dilution tunnel. A secondary dilution tunnel is coupled to the main tunnel, for particulate collection (6). The main tunnel incorporates the principle of critical flow with four ventures to provide a constant mass flow of the dilute exhaust. Different driving test cycles including the heavy-duty chassis cycle (HDCC), CBD, arterial cycle, commuter cycle, and New York Composite Cycle (NYCC) have been integrated into the control system. The CBD cycle has been selected for testing both buses and trucks. The cycle has 14 identical ramps; each takes 40 seconds. This includes 10 seconds acceleration starting from idling, 18.5 seconds constant speed at 32.2 km/hr (20 miles/hr), 4.5 seconds for deceleration to idle speed, and 7 seconds idling time (Figure 2). The outer wheels of the tested vehicle driving axle are removed and the vehicle is driven back, mounted onto the test bed, and chained to the dynamometer. Before starting the test, the vehicle is kept running to warm up the whole system until the differentials of the dynamometer reach 40 °C (100 °F).

At the test site, the laboratory undergoes different check and calibration procedures to ensure that the chassis dynamometer components and the exhaust gas analyzers are functioning properly. The instructions in code 40 (6) and the industrial recommended practice for heavy duty chassis testing are followed during these procedures. Calibration curves are drawn up for each analyzer. The flame of the Flame Ionization Detector (FID) is peaked, the NOx efficiency test is applied on the chemiluminecent analyzer, and the Non-dispersive Infrared Analyzers (NDIR) are checked for water interference.

The calibration procedures are applied to the gas analyzers by using a gas divider with an accuracy of 0.5 %, in conjunction with high quality span gases. The traceability of these gases is 1 % as required by the National Bureau of Standards (NBS). The span gas is mixed with the carrier gas (zero gas) by the gas divider in ratios from 0 to 100 % in an increment of 10 % to calibrate the analyzer. Ten points are generated for each analyzer to draw the calibration curve. A third degree fit equation is generated for the HC, CO and CO_2 analyzers (FID, NDIR), and a second degree fit is generated for the NO_x analyzer (Chemiluminecent). After each run, the analyzers are zeroed and spanned. If the drift is more than 2 % the run is canceled and the analyzer is recalibrated. The dilution tunnel is also calibrated by the propane injection technique to check the mass flow rate of the diluted exhaust and to ensure proper mixing.

A coastdown test is conducted on the chassis dynamometer to determine the system losses, then these losses are subtracted from the power absorbed by the power absorber. The flywheel sets are

Figure 1: Tested Bus Setting on the Chassis Dynamometer

adjusted to simulate the inertia of the tested vehicle. Inertia weight of a vehicle is calculated, using the Gross Vehicle Weigh (GVW) value listed by the manufacturer, as follows:

$$Bus \ Inertia \ Weight = GVW \ (lb) - \frac{1}{2} \ (Seating \ + \ Standing)$$
$$* \ 150 \ lb \ + \ 150 \ lb$$

$$Truck \ Inertia \ Weight = 0.7 \ * \ GVW$$

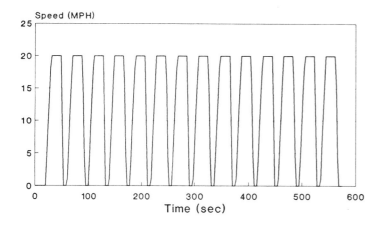

Figure 2: The Central Business District (CBD) Cycle

EVALUATION METHODOLOGY

Tests being performed by the WVU Transportable Laboratory are different from those of the other test facilities since the WVU laboratory tests the vehicles without checking or tuning their engines. In other words, the vehicle is tested under real conditions of field operation. The reasons for this choice are:

1. Some of the tested vehicles use new dedicated alternative fuel engines which are under the supervision of the engine manufacturers, rather than the fleet owner, since these engines are part of pilot experimental projects.

2. A vehicle is available for only one day to minimize fleet operation interruptions.

3. Thorough investigation of vehicle performance is both expensive and time consuming.

The above testing conditions present uncontrolled factors, and do not allow adequate evaluation of the effects of using alternative fuels on exhaust gas emissions. Testing a large number of vehicles is necessary to compensate for the errors that may result from the uncontrolled factors. Therefore, the development of a data base in order to evaluate the use of alternative fuels in heavy duty engines is necessary to determine the real advantages of these fuels. Building this data base is the major objective of the WVU Transportable Laboratory.

TEST PLAN

The CBD hot start test cycle is used to test all vehicles. Normally, four runs are conducted on each vehicle and the average of the tests is calculated along with the standard deviations and variation coefficients. Hot soak is applied for twenty minutes between each two successive runs. One driver is assigned to run all the driving tests to ensure consistency in following the test cycle.

Four Houston Metro buses were tested. Description of the four vehicles, including engine make and type of fuels used, are shown in Table 1. An example of the tested vehicle data sheet is shown in Appendix A. The actual driving CBD cycle is monitored and the deviation from the ideal CBD is shown as a driving error, Figure 3. This error can be minimized if the test driver is well trained. Continuous torque, Figure 3, produced by the tested vehicle driving axle at the hub is measured by two torque transducers, one at each side of the vehicle. The figure shows the change in torque during the acceleration, deceleration, top speed, and idle time of the cycle. Road power and chassis dynamometer drive train losses are also monitored and are shown in Figure 3.

Table 1: Test Schedule of the Houston Metro Transit

Test Number	Test Date	Vehicle Make/Year	Engine	Vehicle Number	Fuel
HM2114-D2/LNG	3/3/92 Tuesday	GM 1983	DD 6V-92 PING	2114 Bus	#2 Diesel /LNG
HM2579-D2/LNG	3/5/92 Thursday	IKARUS 1991	DD 6V-92 PING	2579 Bus	#2 Diesel /LNG
HM2579-CD/LNG	3/5/92 Thursday	IKARUS 1991	DD 6V-92 PING	2579 Bus	Clean Diesel /LNG
HM2579-CD/CNG	3/6/92 Friday	IKARUS 1991	DD 6V-92 PING	2579 Bus	Clean Diesel /CNG
HM2579-D2/CNG	3/6/92 Friday	IKARUS 1991	DD 6V-92 PING	2579 Bus	#2 Diesel /CNG
HM1995-D2	3/7/92 Saturday	GMC 1983	Detroit 6V92TA	1995 Bus	#2 Diesel
HM1995-CD	3/7/92 Saturday	GMC 1983	Detroit 6V92TA	1995 Bus	Clean Diesel
HM2701-D2	3/7/92 Saturday	IKARUS 1992	Caterpillar 3176	2701 Bus	#2 Diesel
HM2701-CD	3/7/92 Saturday	IKARUS 1992	Caterpillar 3176	2701 Bus	Clean Diesel

TEST DATA

Figure 4 shows the integrated continuous rates of CO, HC, and NO_x as a function of the 560 sec CBD cycle time. The figure shows that the production rates of HC, CO, and NO_x increase by acceleration and decrease by deceleration, with the maximum production at the highest speed and the minimum production at idle speed. Emission results of the four vehicles running on different types of fuels are shown in Appendix B.

Actual Driving Speed

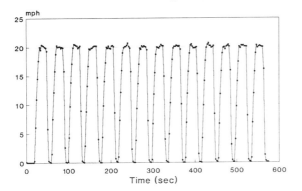

Driving Error

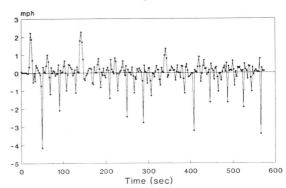

Hub Torques (Side 1)

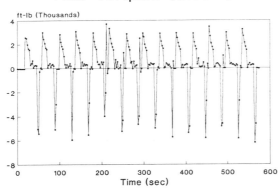

Road-Power and System-Loss

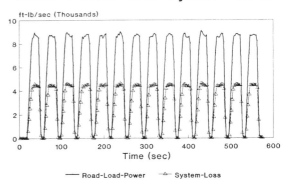

Figure 3: The Integrated Functions of the Chassis Power
Train System During the CBD cycle

Figures 5-8 show the emissions from a 1983 GMC bus powered with a Detroit-Diesel (DDC) 6V-92TA engine. No. 2 diesel and CD were used separately to fuel the bus during the test cycles. Clean diesel produced significantly lower CO, NO_x, and PM rates, while the emission of HC was higher than that of D2.

The second bus, a 1992 Ikarus powered with a 3176 Caterpillar engine, was tested in the same manner. The emission production trends of the two fuels (Figures 9-12) were similar to the trends noticed for the first bus.

Figures 13-16 compare the emission rates of the 1983 GMC and 1992 Ikarus buses. Production of NO_x, HC, and PM from the GMC bus was higher while the CO production was lower than that of the Ikarus bus for the two fuels. To explain the differences between the emissions of the two buses, more detailed investigations are needed.

Two dual-fuel buses were also tested. The first bus was a 1991 Ikarus powered by DD 6V-92 pilot ignition natural gas (PING) engine and was tested with four dual-fuels (D2-LNG, CD-LNG, CD-CNG, and D2-CNG), Table 1. The second bus was 1983 GMC powered with DD 6V-92 PING engine and was tested with D2-LNG. The dual-fuel engine uses the base fuel(CD or D2) during the starting, idling, and partial throttling below 30 %. Once the load increases and the fuel throttling exceeds the 30% ratio, a step-down pressure valve between the gas tank and the combustion chamber opens and LNG or CNG fuel mixes with the base fuel inside the combustion chamber. The first bus was tested for four successive days; each day was assigned to one dual-fuel. Before starting the test the bus was kept running for enough time to purge any residual fuel in the fuel inlet system and the crevices of the combustion chamber from the previous test.

Figures 17-20 show the emission rates of the first bus. Among dual-fuels, D2-CNG produced the lowest levels of CO, HC, and PM; meanwhile it produced the highest NO_x levels. This observation may suggest that CNG was effective in reducing CO and PM in the exhaust since D2, when used as a single fuel, produced higher levels of these two pollutants than CD, Figures 5-12. The results also indicated that a relatively low PM emissions were produced by dual-fuels when compared with D2 fuel. However, much higher rates of HC and CO from all the dual-fuel tests, compared to both single fuels were noticed. This may be attributed to the high production of methane in the exhaust, as this gas constitutes more than 90 % of the CNG or LNG fuels. The three way catalytic converters used in current vehicles can catalyze the oxidation of the non-methane hydrocarbons (NMHC) in the exhaust at temperatures between 500 and 700 °C. Methane oxidation requires higher temperature catalytic converters which are still under research development and are not yet commercially available. Dual fuels produced more or less NO_x levels comparable to the single fuels.

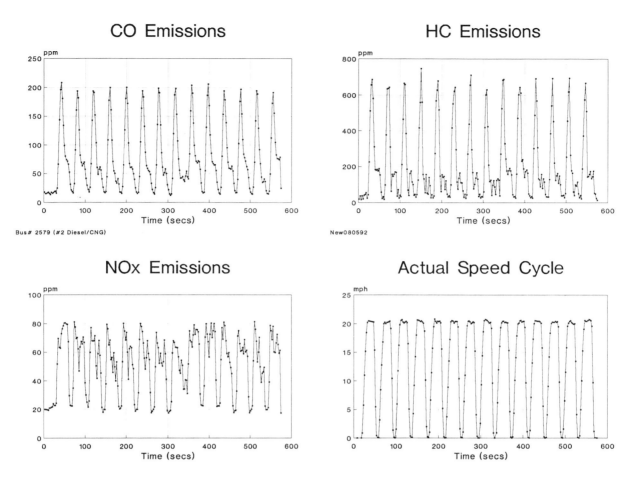

Figure 4: The Integrated Continuous Rates of the Exhaust Emissions

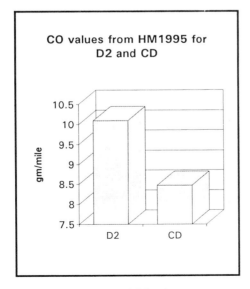

Figure 5: Comparison of CO values
between D2 and CD

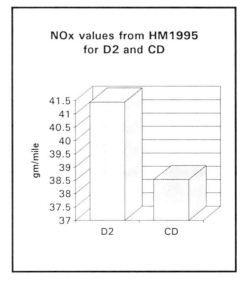

Figure 6: Comparison of NOx values
between D2 and CD

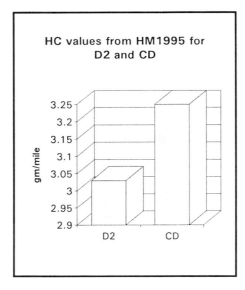

Figure 7: Comparison of HC values
between D2 and CD

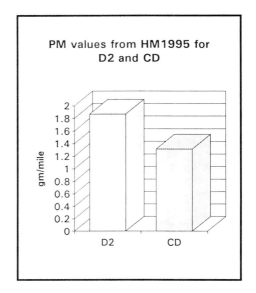

Figure 8: Comparison of PM values
between D2 and CD

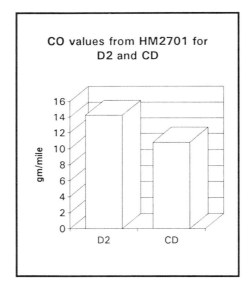

Figure 9: Comparison of CO values
between D2 and CD

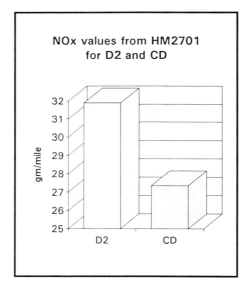

Figure 10: Comparison of NOx values
between D2 and CD

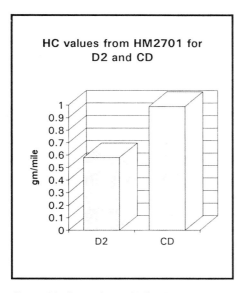

Figure 11: Comparison of HC values
between D2 and CD

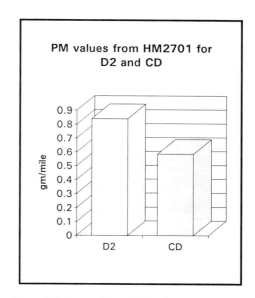

Figure 12: Comparison of PM values
between D2 and CD

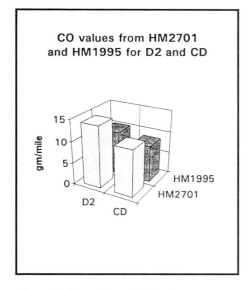

Figure 13: Comparison of CO values
between HM2701 and HM1995

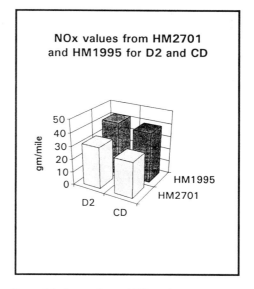

Figure 14: Comparison of NOx values
between HM2701 and HM1995

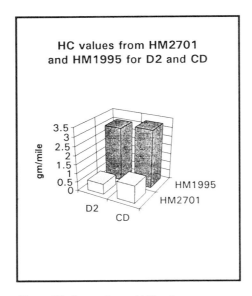

Figure 15: Comparison of HC values
between HM2701 and HM1995

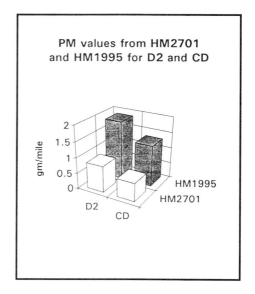

Figure 16: Comparison of PM values
between HM2701 and HM1995

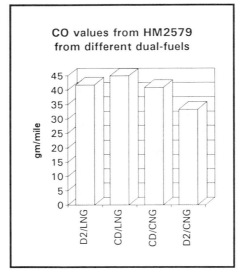

Figure 17: Comparison of CO values
among four different dual-fuels

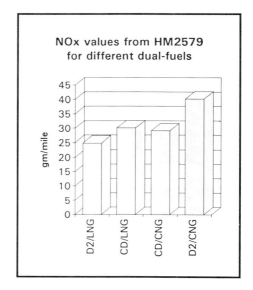

Figure 18: Comparison of NOx values
among four different dual-fuels

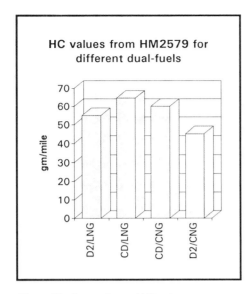

Figure 19: Comparison of HC values among four different dual-fuels

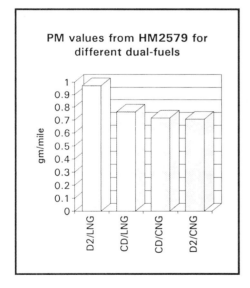

Figure 20: Comparison of PM values among four different dual-fuels

Figures 21-24 show a comparison between the 1983 GMC bus powered with DD 6V-92 PING and the 1991 Ikarus bus powered with a similar engine. Both buses were tested using the dual-fuel D2-LNG. Emission rates of CO and HC from the GM bus were lower than those of the Ikarus, while NO_x and PM rates were higher. Differences in rates of production could be attributed to the differences in GVW, automatic transmission, vehicle make, and/or the efficiency of their catalytic converters.

CONCLUSION

1.   Clean diesel produced less CO, NO_x, and PM, but higher amounts of THC emissions than D2.

2.   Dual-fuels containing CNG or LNG produced slightly less PM than the tested single fuels (CD & D2), while their HC and CO productions were much higher than that of the tested conventional fuels.

3.   Among the tested dual-fuels, D2-CNG produced the lowest CO, HC, and PM emissions and the highest NO_x.

ACKNOWLEDGEMENTS

The authors wish to thank the Department of Energy (DOE), in particular the Office of Transportation Technologies and the Office of Alternative Fuels. A special thanks to A. Chesnes, J. Allsup, J. Russel, and J. Garbak for the grant which made this study possible (contract number DE-FG02-90Ch10451). Thanks are also due to T. Baines and L. Jones, EPA - MVEL, Ann Arbor, MI, for their technical guidance and support.

Important Notice: This project is funded by the DOE and the tests are conducted on vehicles funded by the U.S. Federal Transit Administration (FTA). Tests are managed by West Virginia University. Views mentioned in this paper do not necessarily reflect the views of the above mentioned departments, nor does any mention of any trade names, commercial products, or organizations imply endorsement by the above mentioned departments.

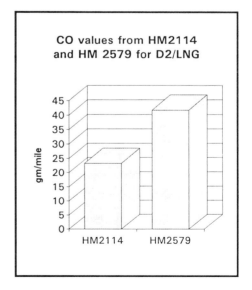

Figure 21: Comparison of CO values between HM2114 and HM2579

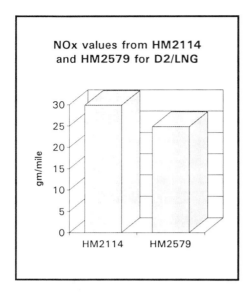

Figure 22: Comparison of NOx values
between HM2114 and HM2579

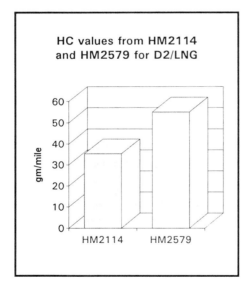

Figure 23: Comparison of HC values
between HM2114 and HM2579

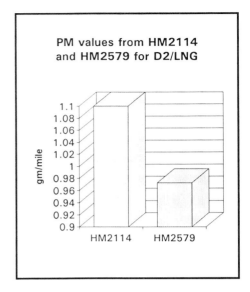

Figure 24: Comparison of PM values
between HM2114 and HM2579

REFERENCES

1.   J. Alson, J. Alder, and T. Barnes, "The Motor Vehicle Emissions Characteristics and Air Quality Impacts of Methanol and Compressed Natural Gas," US EPA, July 1988.

2.   R. Bata, N. Clark, M. Gautam, A. Howell, T. Long, J. Loth, D. Lyons, M. Palmer, B. Rapp, J. Smith, and W. Wang, "The First Transportable Heavy Duty Vehicle Emissions Testing Laboratory," SAE 912668.

3.   R. Bata, N. Clark, M. Gautam, A. Howell, T. Long, J. Loth, D. Lyons, M. Palmer, J. Smith, and W. Wang, "Transportable Emissions Testing," SAE Magazine, Automotive Engineering, January 1992, p. 31-35.

4.   Greg Rideout, "Exhaust Emissions from Natural Gas and Diesel Powered Urban Transit Buses," Environment Canada, December 1991.

5.   Steven Fritz and Ralph Egbuonu, "Emissions from Heavy Duty Trucks Converted to CNG," ASME, 92-ICE-10.

6.   Code of Federal Regulations, Protection of Environment, 40 Parts, 86-99, July 1991.

APPENDIX A

Test Number: HM2701, D2

Agency:	Houston Metro
Contact Person:	L. Luttrell 635-0263
Vehicle Type (Bus/Truck):	Bus HSTN Metro 2701
Vehicle ID#:	1H9416048NA155250
Vehicle Manufacturer:	Ikarus USA
Vehicle's Model Year:	1992
Gross Vehicle Weight Rating (lbs)	35925
Odometer Reading (miles)	10170 (hub)
Transmission Type: Auto/Manual	Automatic
Transmiss. Config.:	ZF
Outside Tire Dia.(in.)	43.
Frontal Area of Vehicle (sq. ft.)	72.25
Tailpipe O.D.(in.)	5.0"
Fuel Tank(s) Capac.(gal.)	120
Number of Axles	2
Engine Type:	Caterpillar 3176
Engine ID#:	416.4-1060-900
Engine Displacement (cu. in.)	10.3 liter
Number of Cylinders	6
Max. Engine Power (hp)	270
Rated Speed:(rpm)	2100
Fuel Type:	#2 Diesel
Oil Type:	40 WGT.
Dynamometer Test Cycle	CBD Cycle

Test Number: HM2579, D2-CNG

Agency:	Houston Metro
Contact Person:	L. Luttrell 635-0263
Vehicle Type (Bus/Truck):	Bus HSTN Metro 2579
Vehicle ID#:	1H9416023MA155217
Vehicle Manufacturer:	IKARUS USA INC
Vehicle's Model Year:	1991
Vehicle Testing Weight (lbs)	35775
Odometer Reading (miles)	8293
Transmission Type: Auto/Manual	Automatic
Transmiss. Config.:	4 spd.
Outside Tire Dia.(in.)	42.25
Frontal Area of Vehicle (sq. ft.)	75.0
Tailpipe O.D.(in.)	3.5"
Fuel Tank(s) Capac.(gal.)	120
Number of Axles	2
Engine Type:	DD6V-92 Pilot Ignition Natural Gas
Engine ID#:	
Engine Displacement (cu. in.)	552
Number of Cylinders	6
Max. Engine Power (hp)	277
Rated Speed:(rpm)	2100
Fuel Type:	#2 Diesel/CNG
Oil Type:	40wt.
Dynamometer Test Cycle	CBD Cycle

APPENDIX B

Emission results of the Four Tested Buses

Test Number: HM2579-D2/LNG

Fuel: **#2 Diesel/LNG**   Engine: **Detroit**   Test Date: 3/5/92
Unit: g/mile

Test Serial No.	CO	NOx	HC	PM	CO2
HM2579-D2/LNG-06	42.03	26.10	55.5	1.03	3083.97
HM2579-D2/LNG-07	41.31	24.84	54.45	0.94	3057.98
HM2579-D2/LNG-08	42.77	24.51	55.89	0.98	3101.17
HM2579-D2/LNG-09	40.76	23.97	54	0.94	3061.84
AVERAGE	41.72	24.86	54.96	0.97	3076.24
Std. Div.	0.87	0.90	0.88	0.04	20.18
CV%	2.09%	3.64%	1.61%	4.39%	0.66%

Test Number: HM2579-CD/LNG

Fuel: **Clean Diesel/LNG**   Engine: **Detroit**   Test Date: 3/5/92
Unit: g/mile

Test Serial No.	CO	NOx	HC	PM	CO2
HM2579-CD/LNG-10	45.19	30.07	63.75	0.76	2913.81
HM2579-CD/LNG-12	45.06	30.69	64.89	0.76	2883.58
HM2579-CD/LNG-13	44.54	29.52	64.92	0.77	2897.14
AVERAGE	44.93	30.09	64.52	0.76	2898.18
Std. Div.	0.34	0.59	0.67	0.01	15.14
CV%	0.77%	1.95%	1.03%	0.76%	0.52%

Test Number: HM2579-CD/CNG

Fuel: **Clean Diesel/CNG**   Engine: **Detroit**   Test Date: 3/6/92
Unit: g/mile

Test Serial No.	CO	NOx	HC	PM	CO2
HM2579-CD/CNG-14	40.52	29.79	62.67	0.52	2800.22
HM2579-CD/CNG-15	39.55	34.52	58.41	0.81	2917.99
HM2579-CD/CNG-16	41.21	21.35	59.1	0.95	2977.46
HM2579-CD/CNG-18	42.15	30.77	59.67	0.58	2883.55
AVERAGE	40.86	29.11	59.96	0.72	2894.81
Std. Div.	1.10	5.56	1.88	0.20	74.03
CV%	2.69%	19.10%	3.13%	28.03%	2.56%

Test Number: HM2579-D2/CNG

Fuel: **#2 Diesel/CNG**   Engine: **Detroit**   Test Date: 3/6/92
Unit: g/mile

Test Serial No.	CO	NOx	HC	PM	CO2
HM2579-D2/CNG-19	33.68	42.37	45.75	0.81	3117.32
HM2579-D2/CNG-20	33.93	40.44	46.29	0.64	3160.58
HM2579-D2/CNG-21	32.27	37.61	43.23	0.67	3061.17
AVERAGE	33.29	40.14	45.09	0.71	3113.02
Std. Div.	0.90	2.39	1.63	0.09	49.84
CV%	2.69%	5.96%	3.62%	12.84%	1.60%

Test Number: HM2114-D2/LNG

Fuel: **#2 Diesel/LNG**   Engine: **Detroit**   Test Date: 3/3/92
Unit: g/mile

Test Serial No.	CO	NOx	— HC	PM	CO2
HM2114-D2/LNG-01	24.56	30.02	34.53	1.26	2576.57
HM2114-D2/LNG-02	25.60	29.78	35.97	1.13	2486.53
HM2114-D2/LNG-03	21.01	29.64	35.58	0.99	2430.19
HM2114-D2/LNG-04	22.14	30.09	34.98	1.02	2469.13
AVERAGE	23.33	29.88	35.27	1.10	2490.61
Std. Div.	2.12	0.21	0.64	0.12	61.96
CV%	9.08%	0.70%	1.81%	11.13%	2.49%

Test Number: HM1995-D2

Fuel: **#2 Diesel**   Engine: **Detroit**   Test Date: 3/7/92
Unit: g/mile

Test Number	CO	NOx	HC	PM	CO2
HM1995-D2-01	9.53	41.49	3.72	1.97	2875.51
HM1995-D2-03	11.75	41.27	2.67	1.43	3241.82
HM1995-D2-04	9.00	41.56	2.70	2.20	2860.91
AVERAGE	10.09	41.44	3.03	1.87	2992.75
Std. Div.	1.46	0.15	0.60	0.40	215.83
CV%	14.45%	0.37%	19.73%	21.17%	7.21%

Test Number: HM1995-CD

Fuel: **Clean Diesel**   Engine: **Detroit**   Test Date: 3/7/92
Unit: g/mile

Test Number	CO	NOx	HC	PM	CO2
HM1995-CD-05	7.40	39.26	3.27	1.36	2735.20
HM1995-CD-06	9.13	38.03	3.24	1.32	2725.53
HM1995-CD-07	8.88	38.31	3.24	1.26	2770.73
AVERAGE	8.47	38.53	3.25	1.31	2743.82
Std. Div.	0.94	0.65	0.02	0.05	23.80
CV%	11.04%	1.67%	0.53%	3.83%	0.87%

Test Number: HM2701-D2

Fuel: **#2 Diesel**   Engine: **Caterpillar 3176**   Test Date: 3/7/92
Unit: g/mile

Test Number	CO	NOx	HC	PM	CO2
HM2701-D2-01	14.71	30.82	0.63	0.76	3165.89
HM2701-D2-02	14.49	30.16	0.51	0.90	3029.65
HM2701-D2-03	13.51	34.70	0.60	0.86	3128.15
AVERAGE	14.24	31.89	0.58	0.84	3107.90
Std. Div.	0.64	2.45	0.06	0.07	70.34
CV%	4.49%	7.69%	10.77%	8.58%	2.26%

Test Number: HM2701-CD

Fuel: **Clean Diesel**   Engine: **Caterpillar 3176**   Test Date: 3/7/92
Unit: g/mile

Test Number	CO	NOx	HC	PM	CO2
HM2701-CD-04	10.67	27.20	1.02	0.61	3124.42
HM2701-CD-05	10.33	27.39	0.96	0.55	3159.97
HM2701-CD-06	11.36	27.44	0.99	0.60	3186.40
AVERAGE	10.79	27.34	0.99	0.59	3156.93
Std. Div.	0.53	0.13	0.03	0.03	31.10
CV%	4.87%	0.46%	3.03%	5.48%	0.99%

AUTHOR INDEX

New Developments in Off-Highway Engines